THE SCENARIO OF POPULATION GROWTH IN INDIA

THE SCENARIO OF POPULATION GROWTH IN INDIA

DHIRENDRA NATH KONAR

AKANSHA PUBLISHING HOUSE
NEW DELHI-110 059 (INDIA)

AKANSHA PUBLISHING HOUSE
R 37-B, Vani Vihar, Uttam Nagar
New Delhi-110059
Email: ektabooks@yahoo.com
Ph. : 9212065579/9811582579

Showroom:
4649/21-B, Ansari Road
Darya Gang,
New Delhi-110002

The Scenario of Population Growth in India

First Edition 2004

ISBN 81–87606–56-8

[The responsibility for the facts stated, conclusions reached etc. is entirely that of the Author. The publisher is not responsible for them, whatsoever.]

PRINTED IN INDIA

Published by M.P. Misra for Akansha Publishing House, New Delhi and Printed at Tarun Offset Press, Delhi.

Preface

The problem of population has been paining me since my childhood which had been passed in rural areas of West Bengal. I have seen how many big families with many children but with limited food had to pass their days very precariously. This pitiable scene has subsequently awaked me to concentrate on the study of population. In fact, my research works are primarily on population and agriculture and my two research degrees viz. M.Phil. and Ph.D. conferred by Calcutta University, had been on agriculture itself. Occasionally, I had written research papers on these two disciplines and published quite a good number of articles in different journals and books in both English and Bengali: Economic Affairs, Economic Studies, Arthaniti, Business Studies, Arthashastra and Artha Beekshan, Occasional Papers of the Department of Economics, Rabindra Bharati University, Conference Volume of the Indian Economic Association and Dhandhanye (Yojana). The present volume is sequel of the articles published there. Obviously, in bringing out this volume, I have tried to supply information as latest as possible and as accurate as possible. So in comparison to the earlier publications there has been a radical change in this volume. I have tried to make the entire study realistic and coherent. In analyzing the issues I have rarely used technical terminology and the analysis has been made in such a manner that even the ordinary readers may get interest in going through it. The book is expected to come to the use of general readers, researchers, administrators, policy makers, teachers and students of all sorts of educational institutions.

In preparing this book, I have got constant inspiration from Dr. Raj Kumar Sen, Professor of Economics, Rabindra Bharati University, Kolkata. My brother, Prof. Birendranath Konar has constantly pressed me for writing a book of this type.

Shubhendu, my elder son duly helped by Dibyendu, his brother, has taken strenuous effort in typing almost the entire manuscript in his computer. Their mother, Supriya, has silently observed the activities of her husband and two sons and supported them with utmost care. Without their help the book could not have been brought out.

Many thanks are due to M/s Akansha Publishing House, New Delhi, particularly to its proprietor Mr. M.P. Mishra who, in spite of remaining physically unknown to me, has taken a lot of trouble to publish the present volume in the shortest possible time. It is hoped that the generous readers will kindly forgive any kind of error detected in the book.

Dhirendra Nath Konar

Contents

Introduction

India is a very big country where live about 16 per cent of the world population. In fact, population is a great threat to us. Besides natural growth rate, there is an influx of population coming to India from various places. As a result, the growth rate of population in India is very high. This high growth rate of population has been eating into the fruits of development that we have been cultivating since attainment of independence in 1947. Naturally, it is desirable to know the exact magnitude of population of India and its States and Union Territories, its growth rate, the sex composition of the people, the age composition, the residence-wise (rural and urban) classification of the people. Unless we have a thorough and first hand information about all such matters we cannot take any concrete decision about any aspect of our growing population. In this book, a very modest attempt has been made to highlight these different aspects of our increasing population. While doing so we have, very often, considered the case of major States our country. However, we have frequently cited the example of West Bengal about which many information (even at the district level) and relevant analysis have been portrayed. Together with this some information about most of the SAARC countries have also been presented.

The entire analysis has been divided into three parts, namely Part A, Part B and Part C. Part A captioning " Human Aspects in India" contains four chapters. The first chapter under the title "Growth of Population in India" has described the census-wise growth of population between 1891 and 2001. We have seen that during the entire period of one hundred and ten years there are four distinct phases of growth of population in India. Besides the actual population figure, the growth rate of population, the estimated or projected population of India in 2011 and the demographic features of the major States of India will be available in this chapter.

A very prominent demographic feature of the whole world is the sex ratio which has gradually been declining over time in most of the developing countries of the world. A clear picture of this in the context of India will be found in Chapter 2 of Part A. It will be seen that though this ratio in India in 2001 has, marginally, improved from that in 1991, it is still far behind the figure of 1901, a hundred years earlier.

Chapter 3 of Part A has been devoted to a detailed study of the problems of child labour in India. A natural consequence of growing population is the prevalence of child labour which is quite evident in different places of India. Besides high growth rate of population, poverty and lack of educational opportunities are the contributory factors leading to a high magnitude of child labour in India. Many children are to pass inhuman lives and most of them simply get withered before reaching the age of twenty. Unless such children are properly taken care of, subsequently they will not be law-abiding citizens of the country and its consequence will be disastrous. It is the duty of all concerned to protect these children form the way of life they have been leading.

Chapter 4 of Part A has dealt entirely with a very significant issue of our society - how to best utilize the services of the senior citizens of our country. The chapter captioned "The elderly in India" portrays not only various aspects of elder people of India but of many parts of the whole world. It is seen that the percentage of aged people all over the world has been increasing. We have explained the basic reasons behind this. However, it is a matter of great pity that in many cases their past services are denied and they are not properly treated too. But "old is gold" says the proverb and we can definitely get a lot from such old persons. It is a matter of great pleasure that the Government of India has recognized the old people and as a mark of respect, has been giving them a bit concessions in some spheres.

Part B of the present work has concentrated on "Urbanization and Related Issues". The first chapter of this part (that is, the fifth chapter of this book) is on the urbanization in India. A growing population cannot stay in a particular place for a long time. Hence with the growth of population, people have sought for changing their place of residence. With it there has been development of urban areas. Urbanization in India and also in Major states has been quickly occurring after independence. People have been leaving the rural areas and have been living in urban areas where they find it more worthwhile to live. In India there are different classes of towns which have, in detail, been discussed in this chapter. Population of a few large cities of our country in the last three censuses has been presented and we have also shown that in 1991 Mumbai and Kolkata were among the twelve large cities of the world. In 2001, Delhi has also become one such large city and these three cities are among the seventeen mega cities of the world, having a population of ten million or more. At the end of this chapter, we have inserted a subtopic "Degree of Urbanization in Bihar and Orissa" where we have fitted the famous Pareto Law, basically used for studying the inequality in income distribution, to measure the degree of urbanization. Problems associated with urbanization have also been explained in this chapter.

There can be no denying the fact that Kolkata is one of the crowded cities of India. Hence the general people will be interested in knowing how in spite of having constant pressure of population, this more than three hundred year old city has been surviving. That is why we have a chapter at the heels of Chapter 5, entitled "Population Pressure and the Problem of Environment of Kolkata" (Chapter 6). In this chapter, we have, to a good extent, analyzed the causes behind the increasing pressure of population in Kolkata. Environmental problems associated with population pressure have also been explained. At the end of the chapter some remedial measures have also been suggested.

The basic thrust of the present volume is to bring into focus the contribution that can be obtained from the so-called weaker people. We have already brought the case of children. Together with them, we wish to pay attention to the invaluable contribution that can come from women and that is why we have included a chapter, Chapter 7 entitled "Women and the Rural Transformation of India", which has, entirely, been devoted to the importance of women in our country. Women can do a lot for the healthy development of the family, of the society and of the country as a whole. It is through the productive advice of the women that we can reduce the present growth rate of population and we can have such a population as will be bearable to our country. Hence what we need is the dissemination of education among the women and the honour and dignity that should be bestowed upon them.

With the growth of urbanization income of most people has been increasing. However, most people are having a monotonous life, that is, they have more or less, a routine work - rise from bed, get ready for going to office, business centers, court, academic institutions and so on, again come back from office etc., go to bed and the like. That means they hardly get time to enjoy themselves. So whenever they get time, they intend to go for a trip, many people even go for outing at the weekend. Thus with the growth of urbanization a very important industry has developed and this is tourism industry. This industry has, of late, become a very growing industry employing thousands and thousands of people, especially women. That is why we have inserted a chapter, the eighth Chapter covering "Unemployment Problem in India and Development of the Tourism Industry". An obvious consequence of high growth rate of population is the generation of unemployment a lot of which can be solved with the healthy development of the tourism industry. We have already seen that numberless children in our country have to live a hazardous life in different ways. Hence to wipe tears from their eyes and also from their mothers' eyes, these children, if they are to remain employed, may be motivated to shift themselves to tourism industry where they are not put at least to torture. Besides, people of all walks of life may make themselves engaged in the tourism industry that has also been giving us sufficient foreign exchange. Considering the problems of the weaker section of people of our society (namely, children, women and the old), this chapter has been added in the present volume.

The last part of the book, that is, Part C has the caption "Regional, Inter-regional and International Aspects of Growth of

Population." At the regional level we have made, more or less, a detailed study of the growth of population in West Bengal and its different districts. At the inter-regional level we have studied various aspects of human development in major States of India. Finally, at the international level, we have made a study on human poverty in India and other SAARC countries.

Chapter 9 of the present work with the title "Growth Of Population in West Bengal" has focused the demographic scenario of West Bengal during the last 100 years between 1901 and 2001. The projected population of this State for 2011 has also been shown here. The decade-wise growth of population of West Bengal at the district level between 1901 and 2001 has also found a place in this chapter. We have seen that in the last decade of the twentieth century though the annual rate of growth of population in this state had, to a good extent, declined, the districts on the border had experienced a higher rate of growth of population. Unless the infiltration is stopped, West Bengal will face serious consequences of explosive population in the coming years.

An extension of Chapter 2 has been done in Chapter 10 under the title "Sex Ratio in West Bengal" where a more detailed analysis of the sex ratio has been made. Here the fundamental reasons behind the fluctuations of this ratio have been explained and we have presented the trend of sex ratio in India and West Bengal with the help of suitable diagrams.

To make our study on the population of West Bengal complete, we have inserted Chapter 11 entitled "In Search of Depressed Areas of West Bengal" where we have shown that here some areas have grown at a rapid pace and have become more developed while some others have progressed at a snail's pace and have, virtually, remained backward. The eighteen districts of this state have been given ranks on the basis of a number of criteria of development in both 1991 and 2001 and necessary conclusion has been drawn.

Various aspects of the growth of population of a country have got a new light with the introduction of the concept of "human development", the chief architect of which was late Professor Mahbub-ul Haq about 15 years ago. We have considered it worthwhile to have a study on human development in India. This has been done in Chapter 12 captioned. "The Inter-State Pattern of Human Development in India" where we have considered various indicators of human development and applied those on fifteen major States of India, having a population of at least ten million in 1991. The emerging conclusions have, obviously, been drawn at the end of this chapter.

The last chapter, that is, Chapter 13, has dwelt on "Human Poverty in India and Other SAARC Countries". By poverty we mean economic poverty which, in turn, may lead to human poverty characterized by a number of indicators. We have mentioned some such indicators and have studied them in the context of SAARC countries, excluding Bhutan and Maldives. We have seen that the prevalence of poverty in the SAARC countries has caused low human development index in the major countries.

Part A: Human Aspects in India

1

GROWTH OF POPULATION IN INDIA

INTRODUCTION

The study of human resources, that is, of the population of a country, is very significant from the point of view of economic welfare of the country itself. It is especially important because human beings constitute not only an instrument of production but also end in themselves. It is necessary to know, in quantitative terms, the number of people living in a country at a particular time, the rate at which the population of a country has been growing over time, the composition of the population, the distribution of population, the level of education, the life expectancy, the sex-ratio and the projected population of the country for future guidance. We have planned to highlight all these aspects in the context of the Indian economy in the following sections.

POPULATION: A GREAT PROBLEM

According to the Holy Bible, the first man in the world was Adam whereas the first woman was Eve, the wife of Adam. Thus the world of human civilization started with only one man and one woman. Then with the passage of time human beings have been increasing at a tremendous rate. Nowadays, the total number of people and the growth of population of the world have become so alarming that the mere existence in this world has simply become frightful. This is particularly true to the third world countries which face a two-fold problem: the number of people has been increasing at a rapid rate while the number of females has been increasing at a lower rate. This is evidenced by the fact that the sex ratio in many such countries has been decreasing over time. The scenario of India is highly relevant in this context.

India, today, possesses about 2.4 percent of the total global land and she has to bear about 16 percent of the world population. In 1901, India had, on record, a population of 238.3 million with population density of 77 per square kilometer. In 1951, the population became 361 million and the density went up to 117 per square kilometer. In 1981, the figure of population rose to 685 million while the density went up to 216 per square kilometer. In the penultimate census, that is, in 1991 our population became 846 million while 267 was the density of population per square kilometer. Finally, in the latest census of 2001 we had a population of 1027 million and there were 324 persons living per square kilometer. Besides, there were about 62 lakhs of people aged 80 plus.

GROWTH RATE OF POPULATION IN INDIA

We shall now study the rate of growth of population in India in different periods starting from 1891 and ending at 2001. To do that let us present the actual population figures of India as recorded in various censuses. Table 1.1 will show this picture.

Table 1.1 shows the population of India in the census years taken between 1891 and 2001. It also indicates the decade-wise absolute increase in population, the decade-wise simple growth rate and also the decade-wise annual exponential rate of growth of population. It can be seen from the table that between 1971 and 1981 both the decade-wise simple growth rate and the annual exponential growth rate, were the maximum. Then both the rates started falling and in the decade between 1991 and 2001 the annual exponential rate of growth of population of India became 1.96 percent (less than two percent) and this is slightly greater than the targeted annual growth rate of 1.8 percent.

For a quick understanding of the nature of growth of population in India between 1891 and 2001 we have used Diagram 1.1 where the height of the bars represents the magnitude of actual population (in million) in India over different censuses. A glance at the diagram reveals that though the hight of the first five bars is more or less same, it starts increasing from the sixth bar. Moreover, the difference between heights of two cosecutive bars becomes really noticeable the moment we reach this bar.

A careful study of the growth of population in India between 1891 and 2001 reveals that there are four distinct phases in its growth pattern. These phases are:

Table 1.1: Growth of Population in India Between 1891 and 2001

Census Years	Population (million)	Absolute Increase (million)	Decade-wise growth rate (percentage)	Annual exponential growth rate (percentage)
1891	236.0	-	-	-
1901	236.0	0	0	0
1911	252.1	16.1	6.82	0.66
1921	251.3	-0.8	-0.32	-0.03
1931	279.0	27.7	11.02	1.05
1941	318.7	39.7	14.23	1.34
1951	361.1	42.4	13.30	1.26
1961	439.1	78.0	21.60	1.97
1971	548.2	109.1	24.85	2.24
1981	683.3	135.1	24.64	2.23
1991	846.3	163.0	23.86	2.16
2001	1027.0	180.7	21.35	1.96

Source: Census of India (various issues) and figures of growth rate are computed from population data supplied therein.

(i) First Phase (between 1891 and 1921): Stagnant growth rate of population.

(ii) Second Phase (between 1921 and 1951): Steady growth rate of population.

(iii) Third Phase (between 1951 and 1981): High growth rate of population.

(iv) Fourth Phase (between 1981 and 2001): High growth rate of population with distinct signs of slowing down.

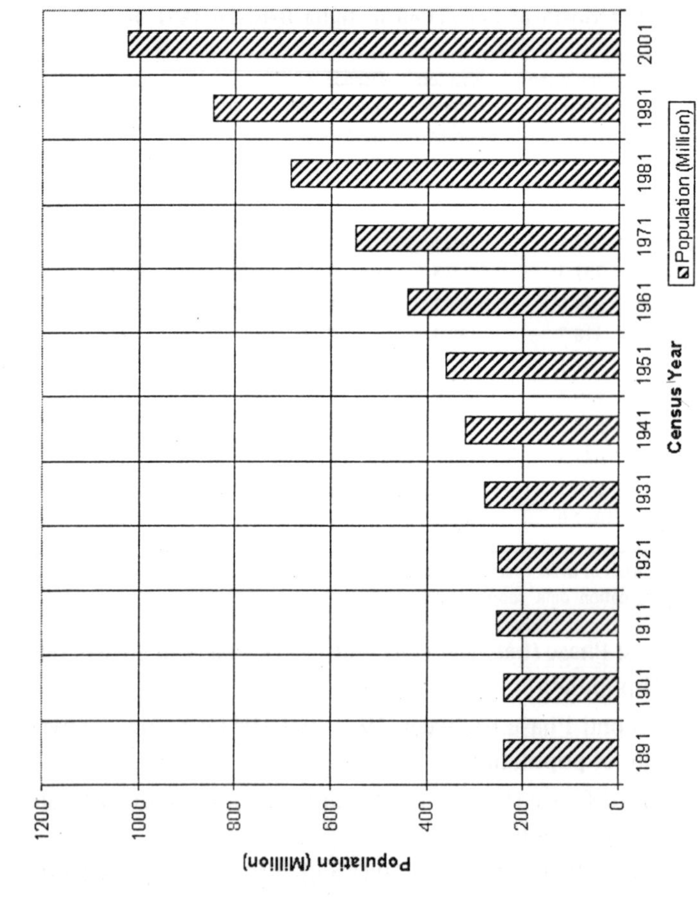

Diagram 1.1: Vertical Bar Diagram Representing Total Population of India

During the first phase of thirty years India's population had grown from 236 million in 1891 to 251.3 million in 1921, that is, just by 15.3 million. In this phase the annual growth rate of population in India was almost negligible, the actual figure being 0.21 percent. The growth of population was held in check by the prevalence of a high death rate which was 42.6 and 47.2 in 1911 and 1921, respectively, against a high birth rate which was 49.2 and 48.1 in 1911 and 1921, respectively. Naturally, in this phase India was in the first stage of demographic transition characterized by a stagnant growth rate of population as has been propounded by Coale and Hoover in *"Population Growth and Economic Development in Low-Income Countries"*.

During the second phase spanning between 1921 and 1951, population of India grew from 251.3 million in 1921 to 361.1 million in 1951, that is, by 109.8 million. The growth rate of population, in this phase, was 1.22 percent compounded annually. Obviously, this growth rate can be considered a moderate rate. The main reason for the increase in population growth in this phase was a sharp decline in death rate from 47.2 per 1000 in 1921 to 27.4 per 1000 in 1951, while there was a very small decrease in birth rate from 48.1 per 1000 in 1921 to 39.9 per 1000 in 1951. During this period India had started its entry into the second stage of demographic transition, which marked a steady but low growth rate of population.

During the third phase, India's population grew from 361.1 million in 1951 to 683.3 million in 1981. Thus there was a record growth of population by 322.2 million in a period of 30 years, yielding an annual compound growth rate of 2.15 percent which was 1.76 times the growth rate of the previous phase. During this phase the death rate declined from 27.4 per 1000 in 1951 to 15.0 per 1000 in 1981 while the birth rate declined from 39.9 per 1000 in 1951 to 37.2 per 1000 in 1981. Comparing the relative falls in these two rates we see that during this phase there was a population explosion in India.

Between 1981 and 2001, India entered the fourth phase of high population growth with definite signs of slowing down. During this period, our population moved from 683.3 million in 1981 to

1027.0 million in 2001—an increase of 343.7 million in a period of just 20 years! In this last phase of twenty years the annual compound rate of growth of population in India was 2.06 percent.

As stated earlier, prior to 1921 India was in the first stage of demographic transition. However, from 1921 onwards it entered the second stage of demographic transition but was not yet in a position to get out of it and enter the third stage of demographic transition characterized by low birth and death rates. The annual growth rate of population between 1991 and 2001, that is, 1.96 percent, is a welcome trend which, if strengthened, can further reduce the growth rate of population. In that case it can be expected that sooner or later India will enter into the third stage of demographic transition.

The natural rate of growth of population of a country depends upon the birth rate and death rate. Consequently, variations in these two rates can provide an explanation for the acceleration of the population growth experienced in India. In the previous paragraphs, we have given some such data. However, the detailed information including that on the expectation of life at birth will be available from Table 1.2 and Diagram 1.2.

Table 1.2: Average Annual Birth Rate, Death Rate and Expectation of Life at Birth in India in Different Years

Years	Births per 1000	Deaths per 1000	Expectation of life at Birth		
			Males	Females	All persons
1901	45.8	44.4	-	-	-
1911	49.2	42.6	22.6	23.3	22.9
1921	48.1	47.2	19.4	20.9	20.0
1931	46.4	36.3	26.9	26.6	26.8
1941	45.2	31.2	32.1	31.4	31.8
1951	39.9	27.4	32.4	31.7	32.1
1961	41.7	22.8	41.9	40.6	41.3
1971	41.2	19.0	46.4	44.7	45.6
1981	37.2	15.0	51.5	50.5	50.5
1991	32.6	11.1	56.8	57.4	57.3
2001	31.0	10.9	63.2	66.8	61.7

Source: Office of the Registrar General and Ministry of Health and Family Welfare, Annual Report (2000-01).

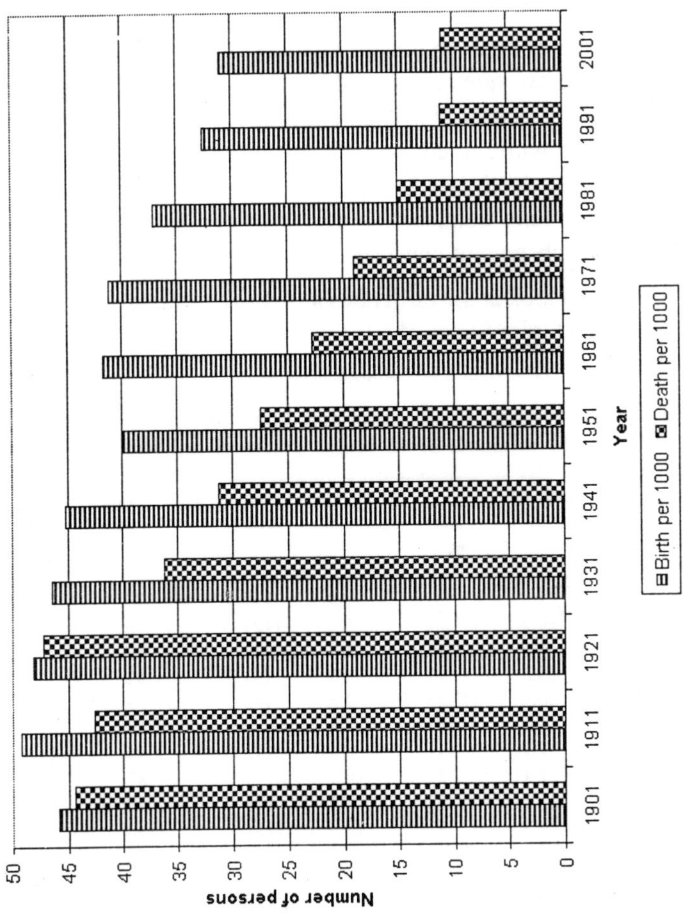

Diagram 1.2: Compound bar diagram representing Birth and death rates in India

We see from the above table that with the passage of time both birth rates and death rates have been declining. However, the rate of decline in death rates, having been greater than that of birth rates, the gap between high birth and falling death rates has been reflected in a high survival rate. Naturally, the high growth rate of population in India can be explained in terms of a persistently high birth rate together with a relatively fast declining death rate.

The nature of birth and death rates can vividly be know if we represent these two rates by a bamd bar chart as shown in Diagram 1.2. From this it is seen that the height of the first and third bars is almost same while in case of the second one the height of the bar representing death rate is distinctly less than that representing the other one. This trend has, gradually, been prominent in the report of the subsequent censuses.

Let us now look at the birth rates and death rates as experienced by some of the major States of our country, which have been shown in Table 1.3.

Table 1.3: Birth Rates, Death Rates and Infant Mortality Rates (and also Ranks attained) of 15 Major States of India in 2000

States	Birth rate per 1000	(Rank)	Death rate per 1000	(Rank)	Infant mortality rate per 1000	Live births (Rank)
Andhra Pradesh	21.3	(5)	8.2	(9)	65	(9)
Assam	26.9	(10.5)	9.6	(12)	75	(11)
Bihar	31.9	(14)	8.8	(11)	62	(7.5)
Gujarat	25.2	(9)	7.5	(5)	62	(7.5)
Haryana	26.9	(10.5)	7.5	(5)	67	(10)
Karnataka	22.0	(7)	7.8	(7)	57	(6)
Kerala	17.9	(1)	6.4	(1)	14	(1)
Madhya Pradesh	31.4	(12.5)	10.3	(13.5)	87	(14)
Maharashtra	21.0	(4)	7.5	(5)	48	(2)
Orissa	24.3	(8)	10.5	(15)	96	(15)
Punjab	21.6	(6)	7.4	(3)	52	(5)
Rajasthan	31.4	(12.5)	8.5	(10)	79	(12)
Tamil Nadu	19.3	(2)	7.9	(8)	51	(3.5)
Uttar Pradesh	32.8	(15)	10.3	(13.5)	83	(13)
West Bengal	20.7	(3)	7.0	(2)	51	(3.5)
All India	25.8		8.5		68	

Source: Ministry of Health and Family Welfare, Government of India.

In terms of the above table, the all-India average figure of birth rate in 2000 was 25.8 and only six states, namely Haryana, Assam, Rajasthan, Madhya Pradesh, Bihar and Uttar Pradesh, had birth rates higher than this figure. It is a matter of surprise that Haryana, occupying the third place in India in terms of per capita income in 1999-2000, has not yet been in a position to reduce the birth rate in a significant manner. In this respect the success of the states occupying the first eight ranks where the birth rate is less than 25, is really creditable and other states should emulate their example.

The all-India average death rate in 2000 was 8.5, which is quite satisfactory. In this respect, the performance of all the states barring Bihar, Assam, Uttar Pradesh, Madhya Pradesh and Orissa is quite good. Coming to the trend of the infant mortality rate which is a very good indicator of economic development of a country, we note that the all-India figure in 2000 was 68, which is quite high when we compare this figure with the figures attained by many developed countries of the world. In this respect we cannot but praise the success achieved by Kerala. Maharashtra, Tamil Nadu, West Bengal, Punjab and Karnataka also had shown commendable success in this affair. However, the performance of Orissa, Madhya Pradesh, Uttar Pradesh, Rajasthan and Assam is very poor. It may be seen that Haryana had just touched the all-India figure of 68 in respect of the infant mortality rate.

We have calculated the Spearman's rank correlation coefficient between birth rate and infant mortality rate and also that between death rate and infant mortality rate, after attaching ranks attained by the above-named states in these three aspects in the usual manner. The former figure has been found to be 0.78 while the latter one has been as high as 0.85. It is likely that the states having high (low) infant mortality rate will have high (low) death rate and our study also corroborates this finding. The birth rate is also highly correlated with the infant mortality rate. Hence one way of reducing the birth rate to a significant extent is to have a spectacular reduction in the infant mortality rate as has been shown by Kerala which has occupied the first rank in respect of each of the demographic features portrayed in Table 1.3. If the parents get assured that their new-born baby will survive, they will, in general, not hesitate to adopt family planning scheme. But if they are not sure of the survival of their new child, they will, unhesitatingly, not adopt family planning programme. Another way of making a steep fall in the birth rate is to make a satisfactory reduction in poverty level. The poor parents living in distant rural areas, seek to have more children because they will, hopefully, try to bring more income in the family having no alternative source of income except that which is earned by the sweat of their brow.

Another very important demographic feature that we have recently been experiencing is the relatively high expectation of life at birth, which can be seen from Table 1.2. In this respect, women are in a better position than men. In 1911 the expectation of life at birth for males was 22.6 years while that for females was 23.3 years. These two figures in 1991 were, respectively, 56.8 and 57.4 years and in 2001 these were 63.2 and 66.8 years, respectively. This is another sign of good development that India has been experiencing. We shall make a detailed discussion about it in a subsequent chapter (that is, Chapter 4).

A very prominent demographic feature revealed through the census report of India is the decline in the proportion of children in the age-group 0-6. According to the Census Commissioner (2001), there is a positive correlation between population growth rate and the child population in the age-group 0-6, which declined from 17.94 percent in 1991 to 15.42 percent in 2001. Such a fall in the proportion of child population to the extent of 2.52 percent is indicative of a fall in fertility level. Table 1.4 shows the proportion of child population in some selected States of India in the last two censuses.

Table 1.4 displays that in Tamil Nadu, Andhra Pradesh, Kerala, Karnataka, Maharashtra, West Bengal and Gujarat, the percentage of child population in the age-group 0-6 is below the national average of 15.42, whereas in States like Haryana, Uttar Pradesh, Madhya Pradesh, Jharkhand, Rajasthan and Meghalaya it is much higher than the national average in 2001.

PROJECTED POPULATION OF INDIA IN 2011

We have already stated that the crucial phase in the growth of population in this vast country started in 1951 as between 1951 and 2001 the population of India increased at the rate of 2.36% compounded annually. The exponential curve (often used for showing the growth of population of a country) representing the population figure of India between 1951 and 2001 has been found to be

$$P_t = 36.495\ (1.236)^{t-1951}.$$

Table 1.4: Percentage of Child Population in the Age Group 0-6 to Total Population in Some States of India in 1991 and 2001

States	1991	2001
Andhra Pradesh	13.33	10.98
Bihar	20.70	19.59
Gujarat	16.48	14.19
Haryana	18.98	15.46
Jharkhand	20.17	17.82
Karnataka	16.63	12.94
Kerala	13.19	11.48
Madhya Pradesh	19.94	17.58
Maharashtra	17.11	13.63
Meghalaya	22.18	19.84
Rajasthan	20.13	18.51
Tamil Nadu	13.33	10.98
Uttar Pradesh	20.38	16.35
West Bengal	16.98	13.88
All India	17.94	15.42

Source: Census of India, Series I, India, Provisional Population Totals.

After fitting this curve we have computed the projected population figures which have been shown in Table 1.5.

We may note here that at the annual growth rate of population of 2.36%, the projected population of India may be around 130 crores when we will hold the next census in 2011. From the fourth column of Table 1.5, we get the idea that in the six decades of the counting of population in free India it was only the third decade when the annual growth rate of population was the least (1.54 %). One interesting feature that also emerges from Table 1.5 is that in the last decade the annual growth rate of population has been less than 2%, the actual figure being 1.96% and this is slightly greater than the targeted annual growth rate of 1.80%.

Table 1.5: Actual and Projected Population of India (Unit: Crore)

Census Years	Population		Decadal growth rate of actual population (%)
	Actual	Projected	
1951	36.1	36.5	-
1961	43.9	45.1	1.98
1971	58.8	55.8	2.97
1981	68.5	68.9	1.54
1991	84.6	85.2	2.13
2001	102.7	105.3	1.96
2011	-	130.2	-

Source: Computed from Census Data.

DEMOGRAPHIC FEATURES OF MAJOR STATES OF INDIA

Let us now dwell on some demographic features of major States of India, each having a population figure of at least 1 crore. These features have been presented in Table 1.6 from which we notice that more than 16% of the country's total population live in Uttar Pradesh which heads the list of nineteen States having more than one crore of people. Next to Uttar Pradesh is Maharashtra which accommodates 9.42% of our total population. There are three newly formed states, namely Chhatisgarh, Jharkhand and Uttaranchal, of which the case of the first two has been considered in Table 1.6 while that of the third one has actual population of 84.79 lakh. Barring Delhi, the density of population is a maximum (904) in West Bengal while Bihar with a density of population of 880 has the second place.

Table 1.6: Some Demographic Features of Major States of India (2001)

States	Total population	%age of total population	Density	Literacy rate		
				All Persons	Male	Female
Uttar Pradesh	166052895	16.17	689	57.4	70.2	43.0
Maharashtra	96752247	9.42	314	77.3	86.3	67.5
Bihar	82878796	8.07	880	47.5	60.3	33.6
West Bengal	80221171	7.81	904	69.2	77.6	60.2
Andhra Pradesh	75727541	7.37	275	61.1	70.8	51.2
Tamil Nadu	62110839	6.05	478	73.5	82.3	64.6
Madhya Pradesh	60385118	5.88	196	64.1	76.8	50.3
Rajasthan	56473122	5.50	165	61.0	76.5	44.3
Karnataka	52733958	5.13	275	67.0	76.3	57.4
Gujarat	50596992	4.93	258	70.0	80.5	58.6
Orissa	36706920	3.57	236	63.6	75.9	51.0
Kerala	31838619	3.10	819	90.9	94.2	87.9
Jharkhand	26909428	2.62	338	54.1	67.9	39.4
Assam	26638407	2.59	340	64.3	71.9	56.0
Punjab	24289296	2.37	482	69.9	75.6	63.6
Haryana	21082989	2.05	477	68.6	79.3	56.3
Chhatisgarh	20795956	2.02	154	65.2	77.9	52.4
Delhi	13782976	1.34	9294	81.8	87.4	75.0
Jammu & Kashmir	10069917	0.98	99	54.5	65.8	41.8
Other States etc.	30968096	3.03	-	-	-	-
Total	1027015247	100.00	324	65.4	75.9	54.2

Source: Census of India, 2001.

Table 1.6 demonstrates that in 2001 the percentage of literate people has been 65.4 while this figure in 1991 was 51.63. Thus there has been a marked improvement in the spread of education in our country. However, the sorry state of affairs is that the percentage of literate females (54.2) is far behind that of literate males (75.9). Naturally, in this sphere we have a long way to go. Among the states of India the highest literacy rate (90.9%) has been attained by Kerala where the gap between the male literacy rate (94.2%) and the female literacy rate (87.9) has been the smallest. Rajasthan has shown the highest gap between the literacy rate attained by males (76.5%) and that attained by females (44.3%). In case of Bihar the literacy rate attained by females (33.6%) is almost half of that attained by males (60.3%). Obviously, women in Bihar are deprived of basic education. The gap between the literacy rates attained by males and females is the same in Bihar and Uttar Pradesh where, however, the literacy rates of males and females are higher by ten percentage points than those attained by Bihar.

India's Population in the Global Perspective

That the population of India is a matter of grave concern can be understood when we look at the absolute figure of population of India or even its annual growth rate *vis-a-vis* that of other populous countries of the world. Table 1.7 will bring out this presentation.

Table 1.7: Population of Ten Most Populous Countries of the World

	Countries	Reference date	Population (million)	Percentage of the world share
1.	China	1.3.2000	1277.6	21.03
2.	India	1.3.2001	1027.0	16.87
3.	U.S.A	April 2000	281.4	4.63
4.	Indonesia	1.7.2000	213.6	3.49
5.	Brazil	1.7.2000	170.1	2.80
6.	Pakistan	1.7.2000	156.5	2.58
7.	Russian Federation	1.7.2000	146.9	2.42
8.	Bangladesh	1.7.2000	129.2	2.13
9.	Japan	1.10.2000	126.9	2.09
10.	Nigeria	1.2.2000	111.5	1.84
Other Countries				40.12

Source: The World Development Report, the World Bank, Various Issues.

The table demonstrates that about 21 percent of the world population live in China and about 17 percent of this population live in India which shares only 2.5 percent of the global land. We have already seen that between 1991 and 2001 the absolute increase in population in India was 180.7 million which is greater than the present total population of Brazil, the fifth most populaous Country of the world.

In the world population has, at present, been increasing each year by 80 million while in India this increase is about 18 million which is the total population of Australia. This is tantamount to saying that with regard to population every year India has been adding one Australia. It is to be noted that the land area of Australia is 2.5 times that of India. Again, between 1991 and 2001 the world population had grown exponentially at an annual rate of 1.4 percent while in case of India this rate has been calculated to be 1.96 percent. These information simply portray the gravity of the problem.

A comparative study of the growth of population between India and Indonesia, the fourth most populous country of the world, reveals that in the decade 1961-71, population had grown in Indonesia at an annual compound rate of 2.08 percent. However, between 1975 and 1998 there was a fall in this rate from 2.1 percent to 1.8 percent. This is attributed to the successful implementa-tion of the family planning programme embarked upon in the 1970s. But it is sad to say that though India is a pioneer in the introduction of family planning programme in the world, she has failed to implement it successfully. Such a failure has resulted in a dismal picture of the all-round development of India because a high growth rate of population in India has eaten into the vitality of the Indian economy.

REFERENCES

1. Census of Indian, Various Issues.
2. Coale, A.J. and Hoover, E.M.: *Population Growth and Economic Development in Low-Income Countries.*
3. Dreze, Jean and Sen, Amartya (1995): *Indian Economic Development and Social Opportunity*, Oxford University Press.
4. Konar, D.N. and Konar, B.N. (1989) : " A Profile of India's Demographic Pattern since 1891", *Economic Affairs*, Vol. 34. Qr. 4.
5. Konar, D.N. (1988): "Changing Scenario of Indian Population: a Projection to 2000 A.D." *Arthashastra*, Vol. 7, No.1
6. Visaria, P. and Visaria, L: "India's Population Scene after 1981 Census," *Economic and Political Weekly*, Vol. XVI, Nos. 44, 45 and 46, Special Number.
7. *The World Development Report*, The World Bank, Various Issues.

2

TREND OF SEX RATIO IN INDIA

A very important demographic feature of any country is the female-male ratio, also called the sex ratio, defined as the number of females per one thousand of males. The sex of a new-born can not be known a-priori, that is, a new-born may be of either sex, a male or a female. The sex of a would-be baby is genetically determined by the sex chromosome of the father's sperm which may be either 'X' or 'Y' with equal probability whereas that of the ovum is always 'X'. If the fertilized ovum consists of the 'XX' sex chromosomes, the baby will be a female and if the sex chromosomes are 'XY', it is a male. So the probability that a new-born will be either one of the sexes should be exactly half.

In reality, however, we get a somewhat different picture almost all over the world. It is commonly observed that in the Western Hemisphere, the trend has been in favour of feminine population whereas in Asia, in general, and in India, in particular, the trend has been congenial to a masculine population. One surprising phenomenon of India's demographic feature is that not only do we have more males than females, but the sex ratio itself has actually been falling (except between the censuses of 1971 and 1981) till the previous census. This trend will be visualized from Table 2.1.

For the diagrammatic presentation of the sex ratio of India the following vertical bar diagram may be consulted.

The diagram reveals that the bar corresponding to year 1901 alone has maximum height while other bars have, gradually, lost their height. Some ups and downs in height have, however, been expressed by the last three bars. All these clearly, demonstrate a declining trend in the sex ratio in India.

Table 2.1: Trend of Sex Ratio in India in all Censuses Between 1901 and 2001

Census Years	Sex Ratio
1901	972
1911	964
1921	955
1931	950
1941	945
1951	946
1961	941
1971	930
1981	934
1991	927
2001	933

Source: Census of India, Various Issues.

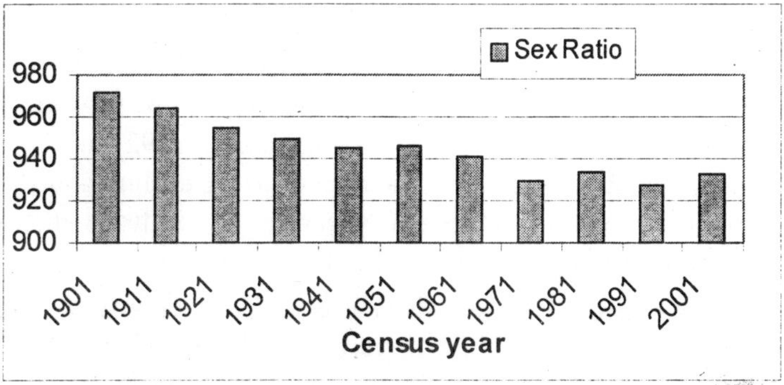

Diagram 2.1: Sex Ratio In India

It is a matter of great surprise to see that the sex ratio in India was the highest (972) in 1901 and this had continuously fallen till the census of 1971. The census of 1981 had shown a slight improvement in this ratio, but in 1991, it, once again, got deteriorated. In 2001, the ratio got improved and had almost touched the ratio attained in 1981. We have seen that in the last eleven censuses the annual exponential growth rate of sex ratio has been -0.43 percent and the exponential curve fitting the sex ratio data has been

$$S_t = 965.7\ (0.9957)^{t - 1901},$$

where S stands for sex ratio and t is the census year

Let us now display the pattern of sex ratio as revealed by each of the thirtyfive States and Union Territories of our country in 1991 and 2001. Table 2.2 has brought this aspect into light. To examine if there is any change in the pattern of sex ratio in the last two census years, we have attached ranks to the States in the usual manner (the case of a tie has also been taken into account). It may be mentioned that in 1991 Kerala alone had a sex ratio (1036) really favourable to women whereas in 2001, besides Kerala, Pondicherry showed such a favourable ratio (1001).

In one decade, both Kerala and Pondicherry had gained 22 points in this ratio. Other States gaining substantial increase in this ratio are Arunachal Pradesh (from 859 to 901), Andaman & Nicobar Islands (from 818 to 846), Uttaranchal (from 936 to 964), Nagaland (from 886 to 909), Uttar Pradesh (from 876 to 898), Meghalaya (from 955 to 975), Manipur (from 958 to 978), Jharkhand (from 922 to 941) and Mizoram (from 921 to 938). But for Daman & Diu there has been an amazing decline in this ratio (from 969 to 709)! Dadra and Nagar Haveli has faced an almost similar situation (from 952 to 811).

There are 24 States where the sex ratio improved in 2001 compared to 1991 while there are 11 States where this ratio declined. All the thirtyfive States and Union Territories of India have been classified according to the nature of female-male ratio in 1991 and 2001 and this has been displayed in Table 2.3.

Table 2.2: Female-Male Ratio in the States and Union Territories of India and Their Ranks in 1991 and 2001

States & Union Territories	1991		2001	
	Ratio	Rank	Ratio	Rank
Andaman and Nicobar Islands	818	34	846	31
Andhra Pradesh	972	6	978	55
Arunachal Pradesh	859	32	901	25
Assam	923	19	932	18
Bihar	907	25	921	21.5
Chandigarh	790	35	773	34
Chhattisgarh	985	2	990	3
Dadra and Nagar Haveli	952	13	811	33
Daman and Diu	969	8	709	35
Delhi	827	33	821	32
Gujarat	934	17.5	921	21.5
Goa	967	9	960	12
Haryana	865	31	861	30
Himachal Pradesh	976	4	970	9
Jammu and Kashmir	896	26	900	26
Jharkhand	922	20	941	15
Karnataka	960	10	964	10.5
Kerala	1036	1	1058	1
Lakshadweep	943	15	947	14
Madhya Pradesh	912	23	920	23
Maharashtra	934	17.5	922	19.5
Manipur	958	11	978	5.5
Meghalaya	955	12	975	7
Mizoram	921	21	938	16
Nagaland	886	27	909	24
Orissa	971	7	972	8
Pondicherry	979	3	1001	2
Punjab	882	28	874	19
Rajasthan	910	24	922	19.5
Sikkim	878	29	875	28
Tamil Nadu	974	5	986	4
Tripura	945	14	950	13
Uttaranchal	936	16	964	10.5
Uttar Pradesh	876	30	898	27
West Bengal	917	22	934	17

Source: Census of India, 2001, Series 20, West Bengal.

Table 2.3: Classification of States and Union Territories of India in 1991 and 2001 According to Nature of Female-Male Ratio

Nature of Sex Ratio (in favour of females)	States and Union Territories	
	1991	2001
(a) Favourable: More than 1000	Kerala. (1)	Kerala and Pondicherry. (2)
(b) Good: More than 950 but less than 1000	Chhattishgarh, Pondicherry, Himachal Pradesh, Tamil Nadu, Andhra Pradesh, Orissa, Daman and Diu, Goa, Karnataka, Manipur, Meghalaya, Dadra and Nagar Haveli. (13)	Chhattishgarh, Tamil Nadu, Andhra Pradesh, Manipur, Meghalaya, Orissa, Himachal Pradesh, Karnataka, Uttaranchal, Goa and Tripura. (11)
(c) Unfavourable: More than 900 but less than 950	Tripura, Lakshadweep, Uttaranchal, Maharashtra, Gujarat, Assam, Jharkhand, Mizoram, West Bengal, Madhya Pradesh, Rajasthan and Bihar. (12)	Tripura, Lakshadweep, Jharkhand, Karnataka, Mizoram, West Bengal, Assam, Sikkim, Manipur, Bihar, Gujarat, Maharashtra, Nagaland, Arunachal Pradesh and Jammu and Kashmir. (15)
(d) Highly Unfavourable: Less than 900	Jammu & Kashmir, Nagaland, Punjab, Sikkim, Uttar Pradesh, Haryana, Arunachal Pradesh, Delhi and Andaman & Nicobar Islands. (9)	Uttar Pradesh, Sikkim, Punjab, Haryana, Andaman & Nicobar Islands, Delhi, Dadra & Nagar Haveli. (7)

Source: Arranged from Census Data.

An alarming situation that has come out from the census of 2001 is a sharply declining gender ratio in the age group 0-6: it decreased to 927 (per 1000 male children) from 945, the level attained in 1991. We may give here such statistics of all the thirtyfive States and Union Territories of India in Table 2.4.

It is obviously seen from Table 2.4 what a surprising situation we are going to face! The decline in the gender ratio is so sharp that even the Provisional Census Report of 2001 questions whether the "sharp decline in the child sex ratios is indicative of any underlying trend of sex-selective abortions."

Table 2.4: Girl-Boy Ratio in the Age Group 0-6 in States and Union Territories of India in 1991 and 2001

States and Union Territories of India	Girl-boy ratio in the age group 0-6	
	1991	2001
Jammu and Kashmir	N. A.	937
Himachal Pradesh	951	897
Punjab	875	793
Chandigarh	899	845
Uttaranchal	948	906
Haryana	879	820
Delhi	915	865
Rajasthan	916	909
Uttar Pradesh	927	916
Bihar	953	938
Sikkim	965	986
Arunachal Pradesh	982	961
Nagaland	993	975
Manipur	974	961
Mizoram	969	971
Tripura	967	975
Meghalaya	986	975
Assam	975	964
West Bengal	967	963
Jharkhand	979	966
Orissa	967	950
Chhattisgarh	984	975
Madhya Pradesh	941	929
Gujarat	928	878
Daman & Diu	958	925
Dadra & Nagar Haveli	1013	973
Maharashtra	946	917
Andhra Pradesh	975	964
Karnataka	960	949
Goa	964	933
Lakshadweep	941	974
Kerala	958	963
Tamil Nadu	948	939
Pondicherry	963	958
Andaman & Nicobar Islands	973	965
All India	945	927

Source: Abridged table from statement 19, pp 5 & 6, Census of India, 2001, Web Edition: Provisional Population Totals, Series I, Paper I of 2001.

From the above picture we may rightly state here that not much has changed since the 1980s when Nobel Laureate Amartya Sen had coined the term "missing women" just to describe those literally not alive because of neglect and discrimination. He had estimated that world-wide there were 100 million missing women. Many believe 50 million, probably more, are in India. After Professor Sen's thesis experts have asserted that women go missing not only because of infanticide and foeticide but also of other reasons. One estimate puts the number of abortions of female foetuses at 3 million to five million a year. A survey of a clinic at Mumbai revealed 7999 out of the 8000 aborted were females. There were more of them missing in the past decade than ever before.

The trend of sex ratio in West Bengal during the last hundred years or so and in the districts (for rural, urban and total population) has been studied, in detail, in Chapter 10. The various reasons behind the fluctuating trend of sex ratio of West Bengal and also of India have been studied. Some references of the growth of population in India, in West Bengal and sex ratio have also been given there.

3

PROBLEMS OF CHILD LABOUR IN INDIA

> "In the little world in which children have their existence whosoever brings them up, there is nothing so finely perceived and so finely felt, as injustice. It may be only small injustice that the child can be exposed to; but the child is small and its world is small..."
>
> – Charles Dickens, *Great Expectations*.

INTRODUCTION

Oscar Wilde in his famous short story entitled *The Selfish Giant* has, through the mouth of the Giant himself said, "I have many beautiful flowers in my garden but the children are the most beautiful of them all." It is a fact that children are the blooming flowers of the garden of the society. Naturally, it is a duty of the members of the society to protect these flowers from a number of damaging effects and also not to pluck them to satisfy our momentary whims. But reality is different from what the previous sentence means to suggest. In fact, children themselves have been treated as labourers called child labourers and the problem of such labourers has, now-a-days, assumed an alarming proportion all over the world.

Keeping this aspect in mind, an attempt has been made here to dwell on different aspects of child labour prevalent in Indian economy. Our discussion will run in the following manner:

Section 1 deals with the meaning of the term "Child Labour".

Section 2 discusses the theoretical issues relating to child labour.

Section 3 analyses the magnitude of the issue of child labour found In India.

Section 4 portrays child labourers found in some towns and cities of India.

Section 5 describes some features of child labourers in India.

Section 6 indicates different causes of prevalence of child labour in Indian Economy.

Section 7, the final one, is the concluding section.

MEANING OF "CHILD LABOUR"

The labour force of any country comprises those persons (excluding the physically handicapped) who belong to the age group 15 to 60 years. Persons less than 15 years of age are not included in this force as they are considered to be too young. Similarly, persons exceeding the upper age limit are also not included in the labour force as they are thought to be too old. Most of the people of the world are of the opinion that children below the age of fifteen, should not be put to strenuous jobs as this is the formative stage of a person's life, when one should pass one's days by studying, playing and gossiping, travelling, merry-making and the like. But because of so many reasons innumerable children, both boys and girls, far less than the prescribed lower age are compelled to come out of their families, to leave their sweet homes, to bid good-bye to their kith and kin and make themselves available for hard work. Naturally, such children are called child labourers. The system of 'child labour' is not at all new. It was prevalent even in old days. However, with the passage of time and with the rapid growth of population this has become a social problem, the rescue from which is called for.

Homer Folks, the Chairman of the United States National Child Labour Committee, defined child labour as "*...any work done by children that interferes with their full physical development, their opportunities for a desirable minimum of education or their needed* *recreation"*[1]. Mr. V. V. Giri[2], the late President of India, distinguished between two senses of the term "child labour". In the first sense, which is used as an economic practice, the term signifies employment of children in gainful occupations with a view to supplementing the labour income of the family. In the second sense child labour is interpreted as a social evil and in assessing the nature and extent of the social evil it is necessary to take into account the character of jobs in which the children are engaged, the dangers to which they are exposed and the opportunities of which they have been deprived. It is mainly in the second sense that the term "child labour" is generally understood. The term "child labour" not only applies to the children working in industries but also to the children working in all forms of non-industrial occupations which are injurious to their physical, mental, moral and social development.

The definition of child labour as used by the International Labour Organization (ILO) and also by the Indian Census Commissioner to estimate the number of working children in India is as follows: A child is classified as a labourer if he (she) is in the age group 5-14 years and is also "economically active". A person will be called economically or gainfully employed if he (she) does work on a regular basis for which he (she) receives remuneration or if such labour results in output for the market. In the Indian context, such a definition of child labour is neither unambiguous nor precise. For example, a child may be working in day to day household chores but is not paid for it and if he (she) may be doing it even at the cost of acquiring education and becoming literate. Moreover, we may look at the number of children in the age group 5-14 years who are categorized neither as child labour nor as students enrolled or attending schools. This part of the child population generally categorized as the "nowhere children" is quite sizeable in India.

THEORETICAL ISSUES RELATING TO CHILD LABOUR

Labour is demanded simply because it has a market called labour market. Similar is the case with child labour which is demanded just because it has a market. Child labour is differentiated from ordinary labour in that labour, in the sense it is usually meant, has a strong bargaining power—there is a trade union for protecting the interest basically of the laboureres (and in the narrower sense of the employers). Trade unions are really strong in the organized sector. In the unorganized sector though there is little trade union, the interests of the labourers (say, of the agricultural labourers) are protected in many ways. But the child labourers have no union. There is none to look after their interests and to raise their grievances. Since child labourers are demanded there is definitely a supply side of it. Needless to say, the supply of child labourers, specially in a highly populated underdeveloped economy like India, is infinitely high, that is, even at a very meagre wage many children will opt for selling their labour. When supply exceeds demand many a thing can naturally happen:

1. Wages are bound to be low or even lower.
2. There are no definite working hours which may vary between 10 hours and 14 hours or even more.
3. There is no security in service.
4. The employment is purely of a temporary nature and the question of permanent absorption by the employer does not simply arise.
5. There will be much exploitation.
6. Buyers' market prevails, i.e., the buyers or the employers have the ball in their court and they can naturally do anything they like.
7. The indifferent attitude of the employers—employers are eager for fulfilling their own interests, i.e., they are desiring to minimize their labour costs by hook or by crook.
8. There is no uniformity in the mode of payment of wages—wages may be paid either in kind or in cash or sometimes even no wage may be paid at all. Sometimes wages may be adjusted for the inherited debt.

In reality, the above facts are found to take place and

educationists, philanthropists, administrators, political leaders and even ordinary people raise a hue and cry for removing child labour from every part of the country. But it has not yet been possible to do so and we are not sure of when child labour will actually be removed from our country. The following sections will touch upon the gravity of the child labour problem prevailing in our economy.

MAGNITUDE OF CHILD LABOUR PROBLEM IN INDIA

It is a matter of great shame to state that all over the world there are about 245 million child labourers while the global population, at present, is about 620 crores. This simply means that approximately 4% of the global population belong to child labourers who have been passing their childhood in a precarious condition even on the threshold of a new millennium. It is a matter of great surprise to say that half of these children are girls who have never known what it means to be a child or to have a few leisurely hours to dream or play. The greatest number of child labourers is found in Asia with a figure of about 44.6 million, followed by Africa with a figure of 23.6 million and Latin America with a figure of 5.1 million. Needless to say, the picture is equally grim for millions and millions of child labourers found here and there in our own land.

It is not exactly possible to give the number of children who have engaged themselves in various activities. We think that no serious attempt has been taken to count the number of such children.

According to the census of 1971, the total number of children (up to the age of 14) was 23 crores 26 lakhs. Of these children about 4.6% or 1 crore 7 lakhs engaged themselves in diverse activities. In the census of 1981 the total number of children was found to be 26 crores 15 lakhs, of whom 5.2% or 1 crore 36 lakhs were child labourers. Of these child labourers about 1 crore 16 lakhs were engaged in agriculture itself, more than 10 lakhs were engaged in construction work, trade and commerce, transport and household work. This information reveals that the highest concentration of child labour is found in the primary sector which till now contributes 29% of our gross domestic product (GDP). However, according to the NSS data, the number of child labourers in our country in 1972-73 and in 1983 was 1 crore 63 lakhs and 1 crore 74 lakhs, respectively. Obviously, there are some differences between the census data and the NSS data. One reason behind it may be that these data relate to varying time periods. According to the 1991 census, the total number of children in our country had increased to 30 crores 72 lakhs. Assuming that about 5% of these children are labourers their number is going to be a gigantic figure of about 1 crore 54 lakhs! Some say that there are about 60 million child labourers in our country. It is also reported that in India every third child has been caught in the web of misfortune, sacrificing the most precious time of their lives, working in grim conditions for minimum and at times non-existent wages.

Let us now bring out the State-wise distribution of child labourers in India as portrayed in Table 3.1.

Table 3.1: State-wise distribution of child labourers in 1981 (in percentage)

States	Locality		
	Urban Area	Rural Area	Total
Andhra Pradesh	13.86	15.20	14.50
North Pradesh	12.35	12.32	12.34
Tamil Nadu	13.29	6.10	9.70
Karnataka	11.08	7.23	9.15
Maharashtra	5.85	10.64	7.83
Bihar	5.25	10.20	7.70
West Bengal	6.76	7.46	7.11
Gujarat	5.36	4.66	5.01
Rajasthan	3.77	5.50	4.63
Orissa	2.18	4.65	3.38
Punjab	2.80	2.11	2.46
Delhi	1.82	0.03	0.93
Other states & Union Territories	15.63	13.90	15.26
Total	100.00	100.00	100.00

Source: UNICEF Publication titled "An Analysis of the Situation of Children in India"(1984).

It is seen from the above table that the highest concentration of child labour is found in Andhra Pradesh which is followed by Uttar Pradesh. In this respect Tamil Nadu has the third place while Karnataka occupies the fourth place. It is also to be noted that Maharashtra and Bihar have almost the same percentage of children engaged in different activities. The place of West Bengal is also worth mentioning. The least concentration of child labour is noticed in Punjab, rich in agriculture. West Bengal, though it has become very rich in the field of agriculture, thanks to land reforms, is still far behind Punjab so far as employment of children is concerned.

According to the Union Labour Secretary, there were about 25 million child labourers in 1998 in our country and the government accepts that about 2 million of them were engaged in hazardous jobs.

A short account of children working in India can be presented in Table 3.2.

Table 3.2: Children working in India in 1991(in million)

Areas	Main workers*			Marginal workers			Total		
	Boys	Girls	Total	Boys	Girls	Total	Boys	Girls	Total
Rural	4.96	3.17	8.13	0.50	1.63	2.13	5.46	4.80	10.26
Urban	0.70	0.25	0.95	0.03	0.05	0.08	0.73	0.30	1.03
Total	5.66	3.42	9.08	0.53	1.68	2.21	6.19	5.10	11.29

*Main workers are those who work for more than 183 days in a calendar year.
Source: Man Power Profile of India, 1997.

We see from Table 3.2 the scenario of children working in rural India: in 1991 out of 11.29 million working children in India as many as 10.26 million, that is, 90.88% had been working in rural areas.

As a supplement to Table 3.1 let us now take Table 3.3 which will reflect the percentage of working children in many of the states of India in 1981 and 1991.

Table 3.3: Working Children* in the Age Group 5 to 14 in India (Percentage)

States	1981	1991
Andhra Pradesh	13.8	10.0
Arunachal Pradesh	11.3	5.7
Assam	-	5.5
Bihar	5.6	4.0
Gujarat	6.9	5.3
Haryana	5.3	2.6
Himachal Pradesh	8.6	4.6
Jammu & Kashmir	15.4	-
Karnataka	11.2	8.8
Kerala	1.5	0.6
Madhya Pradesh	11.8	8.1
Maharashtra	9.4	5.7
Orissa	9.6	5.9
Punjab	5.2	3.0
Rajasthan	8.4	6.5
Tamil Nadu	8.4	4.8
Uttar Pradesh	4.6	3.8
West Bengal	4.1	4.2
All India	7.6	5.4

*Working children in the age group 5-14 pertain to children working as a ratio of total children in that age group.

Source: National Human Development Report, 2001.

Table 3.3 displays that at the all-India level 7.6% of total children had been working in 1981. However, in 1991, that is, within a gap of ten years, this percentage had fallen to 5.4. But we are not sure of what information we will get from the census records of 2001.

CHILD LABOURERS IN SOME TOWNS AND CITIES OF INDIA

Child labourers are found working almost everywhere starting from the Himalayas in the North and ending at Kanya Kumari in the South. In reality, children have become indispensable in almost every kind of industry, starting from carpet weaving and glass manufacturing to sports-goods industry, mines, slate, lock manufacturing units, fireworks and matches, diamond and gem stone cutting and polishing units, leather tanning, tea gardens, tea stalls, hotels, restaurants and canteens, railway platforms, brick-fields, bus terminuses and household services. In our country the highest concentration of child labour is, however, still in the agricultural sector which provides livelihood to about 70% of our population. We may mention here that a few years back the Indian Council of Child Welfare had made a survey and found that in Delhi alone about one lakh children had been engaged in household activities.

Let us now bring out, very briefly, the status of child labourers in some important places of our country.

Shibkashi

It is a few hours' journey from Madurai, the Temple City of South India. Here work many children in fireworks, match industry and presses. In the neighbourhood of Shibkashi, with a radius of 20 km. children of different ages coming from different villages and waiting at different spots, board lorries between 3 A.M. and 5 A.M. and are brought to these factories between 6 A.M. and 7 A.M. Each lorry carries about 200 children. In the factories they are to work for about 12 hours. After the day's work the children following the track of their arrival can reach their home between 7 P.M. and 9 P.M. About 5000 children work in this town itself.

Ferozabad

It is a town in Agra district of the State of Uttar Pradesh where 99% of the total requirement of bangles and ornaments made of glass in India is produced. According to the report of the Labour Department of the Government of Uttar Pradesh, in this town about 70,000 workers are employed. Of them about 13% are child labourers.

Bangalore

Child labourers abound in this garden city, the capital of the

State of Karnataka. In the periphery of the Globe Engineering Factory of Bangalore, lie many hotels, restaurants and tea stalls where services are mainly done by children of different ages. A dream of many children working here is that they may be appointed as waiters when they grow up and they may earn better wages.

Khurja

In the big potteries of Khurja in the State of Uttar Pradesh work about 5000 children.

Secunderabad

At Secunderabad innumerable children remain in the streets for begging, for collecting rag and for searching for valuable items among the wastes. The girls of minor ages earn living through prostitution.

Jaipur

Jaipur, the capital of Rajasthan, is famous for precious stones. Here employment is available for stone polishing activities, for which children even at the age of six or seven make themselves engaged for getting the training. They need about five years to seven years for completing the training during which they do not get wages regularly. They are occasionally paid a nominal amount so that they may show interest in the job.

Aligarh

At Aligarh, the famous town for lock manufacturing, more than 10000 male children are engaged in making different parts of locks. They are paid a very poor amount as wage and sometimes they are to work even seventeen to eighteen hours at a stretch.

Surat

In the diamond factory at Surat in Gujarat, about one lakh males work. Of these persons more than 15% are child labourers whose age is less than ten years.

Mumbai

In Mumbai, the largest city of India, work the largest number of male children. Only on hotel and restaurants alone work more than 50,000 male children and in the leather factories work more than 20,000 such children.

Besides, in the brass factories of Moradabad, in the factories producing gold or silver threads at Lucknow, in the fishing industry of Kerala, in the weaving industry of Kashmir, in different mines of Meghalaya and in *bidi* manufacturing factories situated in West Bengal, Tamil Nadu and Karnataka are engaged millions and millions of children.

SOME FEATURES OF CHILD LABOURERS IN INDIA

After a careful study and through different enquiries, we have found some salient features of child labourers in our country:

1. The child labour is primarily a case of unorganized sector. Child labourers are almost non-existent in both the large-scale industries and the mines. They are found basically in the unorganized sector, namely, the cottage industries, agricultural work, household work and the like.
2. The maximum number of child labourers engaged here and there are illiterate and most of them did not even set foot on the four boundaries of the educational institutions. Some of them had, however, gone to the school and stayed there for a few years. However, after repeated failures in the classes they had left school and joined the labour force.
3. The employment of a child in a particular profession at an early age does not necessarily mean that he/she when grown up subsequently, will be given higher wages.
4. A very small number of children is found to take apprenticeship for a profession not related to agriculture.
5. From the professional point of view, there is a strong case favouring child labourers. There are some jobs which require such skill as can be shown at the early age of one's life. With advancing of age such skills become more difficult to be picked up.

DIFFERENT CAUSES OF PREVALENCE OF CHILD LABOUR IN INDIAN ECONOMY

There are a number of causes behind the wide prevalence of child labour in our country. Of these causes we may point out and discuss the following:

A. Growth of Population in India

One of the basic reasons behind the wide prevalence of child labour in our country is the very high rate of growth of population. How population has been growing in India and in many states of India has, very briefly, been dealt with in Chapter 1.

A natural consequence of the high rate of growth of population is that people, in search of food, shelter and jobs have been concentrating, principally, in four metropolitan cities where slums have sprung up like mushrooms. This is shown in Table 3.4.

Table 3.4: Slum Dwellers in the Four Metros

City	Slum dwellers as a percentage of total population of the city	
	1991	2001
Mumbai	34.3	34.3
Kolkata	32.9	32.9
Delhi	26.7	26.7
Chennai	28.1	28.1

Source: *The Economic Times*, Kolkata, March 29, 2004.

The content of the above table does not require any explanation. It is clear that about 30 percent of total population of the four metros are slum-dwellers. Because of social, cultural and economic reasons the slum dwellers will have more children and naturally there will be more child labourers among them.

B. Population below the Poverty line

A second reason behind the existence of a high percentage of child labourers in India is poverty. It is a matter of great shock that even after 57 years of independence we have not yet been able to eliminate poverty from our country. In fact, there is no magic that can help us eliminate poverty from our land even in distant future. The information contained in Tables 3.5 and 3.6 can speak of this.

Table 3.6 reveals that the number of people living below the poverty line in both absolute and relative terms, which was very high in 1973-74, has been declining over time in both absolute sense and relative terms. This is equally true to rural areas and urban areas.

Table 3.5: Population of India below the poverty line (in Million)

	1973-74	1987-88	1993-94	1996-97	2001-02
Number of poor (million)					
Rural	261.3	231.9	244.0	210.0	137.5
Urban	60.0	72.2	76.3	63.0	44.5
Total	**321.3**	**304.1**	**320.3**	**273.0**	**182.0**
Poverty Ratio (%)					
Rural	56.4	39.1	37.3	30.5	18.6
Urban	49.0	38.2	32.4	25.6	16.6
Total	54.9	38.9	36.0	29.2	18.0

Source: Up to 1993-94, revised data based on the Expert Group headed by Dr. D. Lakdawala, Planning Commission Projections, as accepted by the United Front Government in 1997. Poverty Ratios for 1996-97 and 2001-02 are based on Planning Commission's projections, while the number of poor has been estimated on the basis of these ratios.

We now like to bring out the picture of the population below the poverty line across a number of states of our country. This has been focused in Table 3.6.

It is seen from Table 3.6 that according to the figures of 1993-94 only a few of our states headed by Punjab have less than a quarter of their population living below the poverty line. A natural consequence of such poverty is that these people are ever busy with the problem of filling their stomach. It is not possible for them to bother about their children's education and the like. Quite likely, child labour will be common among these people.

C. Lack of Education

Lack of education and educational opportunities in our country is another reason behind the prevalence of child labour. It is really a matter of great shame that we have not yet been able to make primary education compulsory. As a result of it and also because of massive poverty, the children make themselves available for employment at an early age.

Table 3.6: Number and Percentage of Population Below Poverty Line in Different States of India

States	1983		1993-94		1999-2000	
	No. of persons (million)	Percentage of total population	No. of persons (million)	Percentage of total population	No. of persons (million)	Percentage of total population
Andhra Pradesh	16.5	28.9	15.4	22.2	11.9	15.8
Arunachal Pradesh	0.3	40.9	0.4	39.3	0.4	33.5
Assam	7.8	40.5	9.6	40.9	9.5	36.1
Bihar	46.2	62.2	49.3	55.0	42.6	42.6
Gujarat	11.8	32.8	10.5	24.2	6.8	14.1
Haryana	3.0	21.4	4.4	25.0	1.7	8.7
Himachal Pradesh	0.7	16.4	1.6	28.4	0.5	7.6
Jammu & Kashmir	1.6	24.2	2.1	25.2	0.3	3.5
Karnataka	15.0	38.2	15.6	33.2	10.4	20.0
Kerala	10.7	40.4	7.6	25.4	4.1	12.7
Madhya Pradesh	27.8	49.8	29.9	42.5	29.9	37.4
Maharashtra	29.1	43.4	30.5	36.9	22.8	25.0
Orissa	18.1	65.3	16.1	48.6	16.9	47.1
Punjab	2.9	16.2	2.5	11.8	1.4	6.2
Rajasthan	12.7	3.4	12.8	27.4	8.2	15.3
Tamil Nadu	26.0	51.7	20.2	35.0	13.5	21.1
Uttar Pradesh	55.7	47.1	60.4	40.8	53.0	31.1
West Bengal	31.9	54.8	25.5	35.7	21.3	27.0
All India	322.9	44.5	320.4	36.0	260.2	26.1

Source: Planning Commission, Govt. of India.

It is reported that in our country the number of children who do not get enrolled in any educational institution is at present 7.4 crores. But the number of child labourers is estimated, at present, to be 1.7 crores. The obvious question is: What is the identification of the remaining 5.7 crores of children? There is no positive answer to this question. Naturally, a significant member of these children are girls, who are sincerely engaged in household jobs even at an early age. They are to do lots of jobs almost silently for the betterment of their families—they wash, cleanse, cook, look after their brothers and sisters, carry fuel wood from nearby places, tender domestic animals, bring water from here and there, collect materials from shops and markets and even do services at others' houses. Actually, their services are not taken into account. They should be called child labourers. If that is done the number of child labourers will be greater than what is actually reported.

Again, it is the acute poverty that debars many a child from taking part in formal education even though it is free. Many children (especially girls) born in slums and very poor families, can not go to educational institutions simply because they fail to get the minimum garment to cover their body with. Besides, lack of schools in many villages, lack of accommodation in many schools, discriminatory attitude among friends, economic and cultural diversification and the like are sufficient to give to the sentiment of the children. Its obvious effect is large-scale dropout. In this connection we may point out what has happened in West Bengal.

According to a report of the Education Ministry of the Centre, the number of students from Class I to Class V in West Bengal decreased to 89,07,736 in 1997-98 from 1,01,17,000 in 1996-97. The magnitude of reduction in the number of male students was 4,39,312 while that in the number of female student was 7,69,952. So far as dropouts are concerned the position of West Bengal is second, the first position being taken by Bihar. It is also seen that as the number of closed mills and factories increases and as the opportunity for employment gets shirked, the number of students in schools diminishes. A significant number of the drop-out students are becoming child labourers.

D. Large-scale Unemployment

According to the census of 1991, one-third of the population of our country is estimated to be without work. Such a massive figure of unemployment induces many parents to go in search of any kind of job that they may provide to their children even at an early age. Besides, due to sickness in different industries, closure of mills and lockouts in different factories, many parents have themselves been unemployed. They are to fall back upon their children's income, however meagre it may be. As a result children, even reluctantly, are compelled to take jobs at whatever place and at whatever ways it becomes possible.

E. Miscellaneous

Because of the existence of so many restrictions on employing child labourers in varied types of cottage industries and the prohibition of employment of them in large scale industries, children seek employment in small factories, tea stalls and many other inferior professions even at a nominal wage. Child labour is concentrated primarily in the unorganized sector which is running in its own way. There is no union and there is none to raise the voice of the children who by the sweat of their brow and by bearing all sorts of exploitation from their employers are to be satisfied with whatever they earn.

CONCLUSION

It is a fact that the present children of any country are really the future of the nation in the sense that there are many players, singers, poets, writers, litterateurs and statesmen in themselves. Hence instead of exploiting the children they should be duly brought up in a congenial atmosphere. To make the people of all walks of life aware of this simple phenomenon, each year the 14th day of November (the birthday of Chacha Nehru, our first Prime Minister) since 1957 has been being celebrated as the "Children's Day" in our country. Besides, the first day of June since 1950 has been regarded as the International Children's Day. Moreover, the year 1990 was declared the "International Children's Year". In spite of all these, there is continuous exploitation and abuse of children at various levels and the number of child labourers is also going up year after year. What is the way out? Some suggest that it is through the appropriate legislation that the problem of child labour may be solved. But that is not going to solve the problem, as it is the poverty that has compelled thousands and thousands of children to work here and there. If employment of child labourers gets barred through legal action, many families will simply die of hunger. We should bear in mind that to eliminate child labour poverty should be eliminated first. Hence before attacking child labour we should try our best to attack poverty and at the same time to reduce the growth rate of population through a suitable and effective population policy. If poverty gets addressed, the need for child labour would automatically diminish.

NOTES

1. Quoted in Stein, Emanuel & Davis (ed): Labour Problems in America, Jerome PP 112-113.
2. V. V. Giri: Labour Problems in Indian Industry (page 360)

REFERENCES

1. *Census of India*: Various Issues.
2. *Economic Survey*, Government of India: Various Issues.
3. Giri, V. V. (1958): *Labour Problems in Indian Industry*, Asia Publishing House.
4. *Human Development in South Asia* (1997).
5. Kulshreshtha, Jinesh Chandra (1978): *Child Labor in India*, Ashish Publishing House, New Delhi.
6. Konar, D. N. (1993): "*Shishu Shramik O Nasta Kaishore*" (in Bengali) in *Dhanadhanye* (*Yojana*), July.
7. Konar, D. N. (1989): "Rural Poverty in India—Its Dimension, Causes and Remedial Measures", *Eastern Sociologist*, Vol. 9.
8. *Man Power Profile of India*, (1997).
9. *Magazine of the ILO* (1996): No. 16, June/July.
10. *National Human Development Report*, 2001.
11. *Statistical Outline of India*, 1998-99.
12. Wilde, Oscar: *The Selfish Giant*.
13. *The World Development Report*: Various Issues.

4

THE ELDERLY IN INDIA

INTRODUCTION

An obvious consequence of the process of demographic transition all over the world is the ageing of population. India is not an exception to this trend. In a number of countries and in most international fora, an elderly has been defined to be a person falling in the age group of 65+. However, in India the "ageing of population" is defined in terms of the proportion of persons having age 60 and over in the total population. India has the second largest number of elderly persons after China.

Until a few decades ago, the issues relating to the elderly was not in the forefront of the development agenda in our country but it is only recently that we, in India, have started taking note of this problem. Until the early 1950s,the pattern of high birth rates accompanied by high death rates kept the proportion of India's population aged 60 and above at a low level. However, since the early 1960s the proportion of elderly persons has been increasing due to a steady decline in mortality rates and the consequent improvement in life expectancy and also due to a decline in the fertility rates.

With more and more people falling in the old age group, concern about population ageing has been quite prevalent in the developed countries. However, interest in the subject of ageing of population, especially in the developing countries, is of very recent origin. We may mention here that at the initiative of Argentina, the International Community first debated the question of ageing at the United Nations in1948. Again, the same issue had been raised by Malta in 1969. Also, in recognizing that longevity was becoming one of the major challenges of the twentieth century, the United Nations convened the World Assembly on ageing in Vienna in 1982. The United Nations' General Assembly endorsed the International Plan of Action on ageing in that very year. In 1990, the General Assembly designated October 1 as the "International Day for the Elderly", later renamed the "International Day of the Older Persons". The United Nations' General Assembly decided in 1992 to observe the "International Year of Older Persons" in 1999 to raise awareness of the fast demographic picture of older persons. Accordingly, the year 1999 had been observed all over the world as the year for paying honour, respect and gratitude to the older persons. The theme of the year was *"Towards a Society For All Ages"*. That year had been observed as a means for stimulating debate, promoting action strategies and encouraging research and information exchange.

In the World Assembly at Vienna there was the Indian official delegate who stated that our country did not have much of a problem of the aged people. This is because of the fact that the traditional family structure included the prevalence of joint family system and the significant role of the elderly in taking decision at the household level ensured that most of the elderly in the society were taken care of by the members of their respective families themselves and the government health services took care of their health needs. The interest in the subject of ageing of population has since increased. This is particularly significant for a country like ours where a vital interest so long has been related to the controlling of the size of the population.

THE SCENARIO OF AGED PEOPLE ACROSS THE WORLD

A meticulous study of age distribution of people around the world will reveal that the more and more people have been falling in old age group with age sixty plus. This is tantamount to saying that the life expectancy of people has been increasing. According to a report prepared on the basis of the statistics of fifty countries of the world published in the "Span" magazine in 1992, the actual number of aged people in that year was 34,20,00,000 which was about 6.2% of the world population. Across the world, the number of such people has, in every month, been increasing by more than eight lakhs. The demographers think that by 2010 this number will rise to eleven lakhs. According to this report about 62% of the aged people of the globe are inhabitants of the developing countries. This report also shows that between 1990 and 2005 the percentage increase in the aged people will be maximum (414) in Indonesia the fourth highly populated country of the world and a cub of Asia, alone whereas in case of India this percentage will be 242. The corresponding figures for America, England and Sweden will respectively be 89, 40 and 33. Again, 40 percent of the people having age 80+ plus lived in China, America and Erstwhile Soviet Russia. Moreover, in the world, the Japanese people have the longest expectation of life (81) in 1999. According to the latest information (vide *Aajkaal,* a Bengali daily dated 2.8.2002), the average life expectation of a Japanese woman has been 85.26 years and that of a man has become 78.42 years. Besides, the number of aged women in the world has exceeded that of the aged males. At the same time among the aged people of the world, the growth rate has been maximum for those having age of 80 plus.

According to an estimate made by the United Nations in1980, 5.9 percent of the total population of the world comprised the aged people but in 2005 this percentage will move to 9.7. This figure in the Asian countries was 4.4 in 1980 while by 2025 it will proceed to 10. Besides, in 2025 the number of aged people in Asian countries alone will be 49 percent of the total aged people of the world. It is also reported that by 2000, 83 percent of the aged people of the world will live in the under-developed and developing countries, 6.1 percent of total population of Asia will be aged and 13.3 percent of the people of the developed countries of the world will be aged.

We shall now bring into light the actual number of aged people of some countries of the world. Table 4.1 will convey this information.

Table 4.1: Actual Number of Aged People in Some Countries of the World

(Unit: Crore)

Countries	Years			
	1950	1980	1990	2000 (estimated)
China	4.10	7.30	10.10	13.20
India	2.00	4.20	5.90	7.60
Other regions and Islands of Asia	1.70	3.10	4.30	5.80
Latin America and the Caribbean Islands	0.90	2.10	3.10	4.30
Sub-Sahara Africa	0.90	1.70	2.30	3.20

Source: Human Development Report, 1993.

The table will speak of itself about the increasing number of aged people in many countries of the world. Thus between 1950 and 1980 the number of aged people in India had increased by more than 110.0 percent, in case of China this increase had been about 78 percent. In Latin America and Caribbean Islands this increase had been about 133 percent and for Sub-Sahara this rate had been about 88 percent.

Table 4.2 will present the proportion of aged people in the total population in the SAARC countries.

Table 4.2: Proportion of Aged People in Total Population in the SAARC Countries

Countries	Proportion of aged people as percentage of total population	
	1998	2015 (estimated)
India	4.8	6.4
Pakistan	3.1	3.8
Bhutan	4.0	4.4
Nepal	3.6	4.1
Bangladesh	3.2	4.3
Sri Lanka	6.4	9.3

Source: Human Development Report, 2000.

Table 4.2 demonstrates that in 1998 in Sri Lanka 6.4 percent of the total population had fallen in the old age group whereas in India this rate had been 4.8 percent. We have failed to collect necessary information about Maldives. For other countries in the SAARC group the table itself will speak the fact.

We shall now present some countries (which are highly developed having high human development index) of the world where the proportion of aged people in total population is at least 10 percent. This presentation has been done in Table 4.3.

Table 4.3: Proportion of Aged People in Total Population in Some Highly Developed Countries

Countries	Population and Ranks*	
	1998	2015 (estimated)
Italy	17.6 (1)	22.6 (2)
Japan	16.1 (2)	24.6 (1)
United Kingdom	16.0 (3)	18.7 (4)
Germany	15.9 (4)	20.3 (3)
France	15.6 (5)	18.4 (5)
Malayasia	15.4 (6)	18.4 (5)
USA	12.5 (7)	14.7 (8)
Canada	12.5 (7)	16.2 (6)
Australia	12.1 (9)	15.2 (7)
Poland	11.6 (10)	14.3 (9)
New Zealand	11.6 (10)	14.1 (10)

* Figures within parentheses indicate rank attained by the country concerned.
Source: Arranged from Human Development Report, 2000.

We notice from Table 4.3 that in 1998 the highest proportion of the aged people had been attained in Italy. But the estimated proportion in 2015 reveals that in 2015 Italy will be replaced by Japan where according to the earlier information by 2020, 25 percent of the Japanese people will be aged. In many countries the ranks in 2015 have simply been inter-changed with those attained in 1998. The Spearman's rank correlation coefficient between ranks attained by the above-noted eleven highly developed countries of the world, has been found to be +0.94. This indicates that there is a very nominal change in the ranks attained by the countries in 2015 in comparison to what they had done in 1998.

We shall now focus on the life expectancy at birth and infant mortality rates across countries of the world so as to get a general trend of longevity of the people in this world. Table 4.4 will do so for the countries belonging to the SAARC while Table 4.5 will give the corresponding presentation of some of the very rich countries of the world.

Table 4.4: Life Expectancy and the Infant Mortality Rates in the SAARC Countries

Countries	Life Expectancy		Infant Mortality Rate per 1000 live births	
	1970-75	1999	1970	1998
India	50.3	63	130	69
Pakistan	50.6	63	118	95
Bhutan	43.2	61	156	84
Nepal	43.3	58	156	72
Bangladesh	34.9	61	148	79
Sri Lanka	55.5	73	115	43

Source: Human Development Report, 2000 and World Development Report, 2002.

Table 4.4 displays that within a span of 25 years there has been a significant improvement in the life expectancy of the people in the SAARC countries. There has also been a noticeable fall in the infant mortality rates in these countries. Sri Lanka had the longest life expectancy in 1999 followed by India, Pakistan, Bhutan and Bangladesh. Again, it is Sri Lanka which has reduced its infant mortality rates from 115 in 1970 to just 43 in 1990. The achievement made by all the SAARC countries except Pakistan in the sphere of such rate is really commendable.

A study of Table 4.5 shows that Japan had, in 1999, attained the life expectancy of 81 years. The remaining countries, barring Poland and Malaysia, had performed almost similar attainment. Coming to the vital statistics of the infant mortality rates we see that Japan attained spectacular success in reducing this rate. In this respect every country is almost equally good, though we desire much from Poland and Malaysia.

Table 4.5: Life Expectancy and Infant Mortality Rates in Few Rich Countries

Countries	Life Expectancy		Infant Mortality Rate per 1000 live births	
	1970-75	1999	1970	1998
Italy	72.1	78	30	6
Japan	73.3	81	14	4
United Kingdom	72.0	77	18	6
Germany	71.0	77	22	5
France	72.4	79	18	5
Malaysia	71.3	72	46	9
USA	63.0	77	20	7
Canada	73.2	79	19	6
Australia	71.7	79	17	5
Poland	70.5	73	32	10
New Zealand	71.7	77	17	5

Source: Human Development Report, 2000 and World Development Report, 2002.

We now intend to describe the picture of the aged people in India and some of its major States. The following sections will convey these aspects.

ELDERLY PEOPLE IN INDIA DURING THE LAST HUNDRED YEARS

The proportion of aged persons in total population of our country has been rapidly increasing. This is particularly true since 1961, though in the first half of the last century there were some ups and downs in it. Table 4.6 provides relevant information about the aged persons of India over the last hundred years between 1901 and 2001.

Table 4.6 demonstrates that there were, in India, 12.1 million aged persons in 1901. This figure in 1951, in the first post-independence census, raised to 19.61 million representing an increase of 62.6 percent in 50 years—a simple annual rate of 1.25 percent and

Table 4.6: Elderly (60+) Population, Women and Men, India 1901-2001

Year	Aged as % of total population	Actual number (in million)			Decadal percentage change of population (Aged)			
		Persons	Women	Men	Persons	Women	Men	Total population
1901	5.06	12.06	6.56	5.50	–	–	–	–
1911	5.22	13.17	6.99	6.18	9.20	6.55	12.35	6.82
1921	5.37	13.48	7.00	6.48	2.35	0.21	4.87	-0.32
1931	5.09	14.21	7.27	6.94	5.42	3.74	7.12	11.02
1941	5.66	18.04	9.15	8.89	26.95	25.95	28.03	14.23
1951	5.43	19.61	9.94	9.67	8.70	8.66	8.78	13.30
1961	5.63	24.71	12.35	12.36	26.01	24.28	27.77	21.60
1971	5.97	32.70	15.83	16.87	32.34	28.04	36.56	24.85
1981	6.42	43.98	21.49	22.49	34.50	35.80	33.28	24.64
1991	6.55	55.30	27.07	28.23	25.74	25.96	25.51	23.86
2001	7.70	75.93	37.71	33.22	37.31	39.31	35.41	21.35

Source: Registrar General of India, Censuses of respective years.
For 2001,Projection of the Expert Committee, EPW Vol. XXXIV No. 44,October 30- November 5,1999.

a compound annual rate of 0.98 percent. This figure further increased to 55.30 million at 1991 count, indicating a rise of 182 percent in 40 years, a simple annual rate of 4.55 percent and a compound annual rate of 2.63 percent. Thus between 1951 and 1991 the aged people had grown up at an annual compound rate of 2.63 percent while in the first half of the previous century they had increased at an annual rate of merely 1.00 (0.98) percent.

Comparison of the decadal rate of increase in the aged people with total population of India (that is, columns 6 and 9 of Table 4.6) reveals that except for the decades 1921-31 and 1941-51, the rate was higher for the aged than for the total population. There is the likelihood that the epidemics of 1931 had taken a heavier toll of the aged and there was the equally heavier migration of the young due to partition of the then India, thus reducing the share of the old in the population. It is also clear from columns 4 and 5 of the same table that up to the first six censuses of the twentieth century there were few aged men than aged women. In 1961, the number of aged men was slightly larger than that of women. However, between 1971 and 1991 (the figures for 2001 are not actual figures but projections), there were more aged men in the population than aged women.

SEX RATIO FOR AGED

We may now consider the sex ratio as defined in an earlier chapter, that is, chapter two, for the aged people in our country. Table 4.7 will represent this situation.

We have already seen in chapter two that the sex ratio of Indian population has always been unfavourable to women. Surprisingly, however, this ratio for the aged population is favourable to women though it faces the same declining trend and the figure for 1991 census showed that this ratio for older population was closer to that of general population.

SEX-WISE PERCENTAGE OF AGED PEOPLE ACROSS MAJOR STATES

We now show the share of aged people (men and women separately) in the total population of the major States of our country. Table 4.8 will make it prominent.

Table 4.7: Sex Ratio for Aged In India

Census Years	Sex Ratio
1901	1192
1911	1130
1921	1080
1931	1046
1941	1049
1951	1028
1961	1000
1971	938
1981	960
1991	930

Source: Census of India, Various Issues

Table 4.8: Proportion Of Aged (60+) Population By Sex As A Percentage Of Total Population In 1981 And 1991, In Major States

STATES	1981			1991		
	Persons	Men	Women	Persons	Men	Women
Andhra Pradesh	6.65 (6)	6.48	6.82	6.78 (10)	6.64	6.93
Assam	N.A.	N.A.	N.A.	5.33 (16)	5.65	4.98
Bihar	6.80 (5)	6.77	6.83	6.26 (14)	6.53	5.97
Gujarat	5.33 (15)	5.52	5.16	6.39 (12)	5.96	6.84
Haryana	6.34 (12)	6.87	5.73	7.70 (4)	7.43	8.01
Himachal Pradesh	7.50 (2)	8.31	6.68	8.12 (2)	8.49	7.74
Karnataka	6.62 (7)	6.46	6.79	7.18 (7)	6.80	6.99
Kerala	7.50 (2)	7.15	7.84	8.82 (1)	8.33	9.29
Madhya Pradesh	6.45 (8)	6.09	6.83	6.63 (11)	6.51	6.77
Maharashtra	6.39 (10)	6.07	6.72	6.98 (8)	6.69	7.30
Orissa	6.39 (10)	6.13	6.66	7.20 (6)	7.18	7.23
Punjab	7.80 (1)	8.28	7.25	7.84 (3)	8.07	7.58
Rajasthan	6.03 (13)	5.78	6.29	6.29 (13)	6.08	6.52
Tamil Nadu	6.41 (9)	6.52	6.30	7.45 (5)	7.65	7.25
Uttar Pradesh	6.84 (4)	7.05	6.60	6.86 (9)	7.22	6.46
West Bengal	5.55 (14)	5.32	5.80	6.05 (15)	5.93	6.17
All-India	6.49 (-)	6.40	6.58	6.70 (-)	6.69	6.71

SOURCE: Ageing Population of India—An Analysis of the 1991 Census Data, Table I, Pages 27-34,March1999; Census of India 1981, Series I, Social and Cultural Tables 1987. Figures within parenthesis indicate ranks attained by the States.

In Table 4.8, we have represented the sex-wise proportion of the aged people in the total population of the majority of our States in 1981 and 1991 (we have not yet got the corresponding figures for the latest census of 2001). This Table shows that in1981, Punjab with 7.8 percent of its population falling under the aged (60+) category had the highest percentage of aged population in the country followed both by Kerela and Himachal Pradesh(7.50 percent), Uttar Pradesh(6.84 percent),Bihar(6.80 percent), Andhra Pradesh (6.65 percent) and Karnataka (6.62 percent). All these States had percentages higher than the All-India proportion of 6.49 percent. The results of the next census of 1991 display slight change in the ranks attained by the above-named States. In 1991, 8.82 percent of Kerala's population fell under the aged category, this percent being the highest in the country. Himachal Pradesh with 8.12 percentage of aged and Punjab with 7.84 percentage of aged people, had got the second and third rank respectively. These States were followed by Haryana (7.70 percent), Tamil Nadu (7.45 percent), Orissa (7.20 percent), Karnataka (7.18 percent), Maharashtra (6.98 percent), Uttar Pradesh (6.86 percent) and Andhra Pradesh (6.78 percent). These are the ten States having higher percentages of older people than the All-India proportion of 6.70 percent.

The census results of both 1981 and 1991 indicate that at All-India level the proportion of aged women is marginally higher(by 0.02 percent) in 1991 and a little more(by 0.12 percent) in1981. This pattern is also revealed by the majority of States dealt with here–in 1981 there were 9 States revealing this trend while in1991 there were ten States supporting this pattern. In both the years the number of States not supporting this pattern was six. It may be noted that the percentage of aged women in Kerala is highest in our country. The census records of both 1981 and 1991 support this hypothesis. It may also be noted that Karnataka, Kerala, Madhya Pradesh, Maharashtra, Rajasthan, Andhra Pradesh and Orissa are the seven States where the proportion of aged women is relatively greater than that of men. On the other hand, Himachal Pradesh, Punjab, Tamil Nadu and Uttar Pradesh are four States where the proportion of aged men is relatively greater than that of the aged women.

REASONS BEHIND INCREASING NUMBER OF AGED PEOPLE

We may now talk something about various reasons leading to a high expectation of life, some of which are:

a) Increase in income of many people,

b) Progress of medical sciences,

c) More supply of nutritious food,

d) Increase in a person's awareness of individual health and family health too,

e) Implementation of various schemes for improvement of health,

f) Dissemination of education and

g) Learning through various media.

There can be no denying the fact that an increase in income of the people of a country is the basic indicator of the economic growth of a country. After the Second World War there has been economic stability in many countries of the world and with it the income of those countries has increased. However, with the increase in income there has been discrimination in the distribution of income in the sense that the standard deviation (S. D.) emerging from such distribution has also been increasing and such distribution, far from being moderately symmetrical, has become highly skewed. An increase in income makes people economically better off and they become careful about their physical and mental health.

With the dawn of the twentieth century it has been observed that the expectation of life of people of a country is highly correlated with the per capita income—with the increase in per capita income the expectation of life has also been increasing. For illustration in 1900 the expectation of life of the Americans was only 49 years but that in 1990 has risen to 71 years. With it the infant mortality rate has also significantly come down. In case of China, the expectation of life at birth had increased to 69 years in 1990 from 56 years in 1975 and the infant mortality rate had been reduced to 43 in 1990 from 85 in 1975. Many such illustrations have been submitted in earlier tables.

Other reasons causing higher expectation of life need not be explained. With the progress of civilization, education gets disseminated and both man and society become very much cautious of health.

ARE AGED PERSONS MENTALLY OLD?

It is a fact that human beings are to pass through ten stages in life. Of these stages oldness is an important one and, naturally, we are to cross this stage also. Obviously, ageing gets reflected in our body. In spite of getting aged many persons may remain ever young and so they remain more active and productive—even strenuous effort may not tell upon their health. This will, however, depend upon the attitude of the person concerned towards his life. We may mention here that George Bernard Shaw (1856-1950) and E. M. Foster had remained fully active and alert even at the age of ninety. The internationally reputed litterateur Nirad C. Chaudhuri, remaining fully alert, had died after attaining the age of 101 years! Shri Annada Shankar Roy had celebrated his nineteenth birthday on 15th March 2002. Another stalwart in Bengali Literature Mr. Arun Mitra, had died at the age of ninetythree. Mr. Morarji Desai, our former Prime Minister, had died at the age of ninety-nine. Rabindra Nath Tagore (1861-1941) had remained fully active till the last day of his eternal journey. Professor Prashanta Chandra Mahalanobis (1893-1972), the great Statistician, had been working hard till the last moment of his life and had raised the fame of India in the field of Economics and Statistics all over the world. Thomas Hardy, the great English poet, had been utilizing his pen till the age of eighty for literacy composition. Professor Goodwin, a great Economist, had lived up to the age of ninety-two. Professor Amiya Kumar Dasgupta and Professor Bhabatosh Dutta, the two well-known figures among economists from Bengal, had lived upto the age of eighty-nine and eighty-two respectively. In fine, we may say that Mrs. Amita Sen, the mother of Professor Amartya Sen, the Nobel Laureate, has still been working even at the age of about ninetytwo.

It can be stated that in the present set-up, man has not only been living a longer life but has also been remaining active and productive for a longer period. However, for a quick solution of the unemployment problem the Central Governments, the State Governments and various institutions have been lessening the age of superannuation of the employees and at the same time, have been implementing the Voluntary Retirement Scheme (VRS). But many persons who get retired at an early age or even at the age of sixty, say, have been doing so only at the prime of their experience. Such retired persons are much more interested in performing the works and duties they have been entrusted with, not in getting the retirement benefits of various nature. It is the work and duty which, if performed well, will raise our social status, economic security and identity. On the other hand, the different types of retirement benefits that they get, may snatch away their boastness, their prestige and even independence and they may feel relatively depressed. In macro economic theory, there is a chapter called "National Income Analysis" where it is said that the pension that a retired person gets, may be his personal income but that is not a part and parcel of national income where it is treated as merely 'transfer payments'. In this respect, the mental conditions of the retired persons, being fully aware of themselves, should also be duly honoured. There can be no denying the fact that mental tranquillity and satisfaction that develops with the performance of work and duties strengthens the mind and this inspires a person to work more. If, on the other hand, the person concerned apprehends that his services are going to end, he feels uneasy and such feelings tell upon his physical and mental health. What we want to say here is that many persons, just on the verge of being retired, do not, at once, become inactive and unproductive. To retire from service is only a social rule. Those who feel tired and uneasy before reaching the age of retirement, may have a separate story. We have been talking about those who feel full of vim. What to do with them? To this let us now turn.

PROBLEMS OF OLDER PEOPLE

What emerges from the previous paragraphs is that after the World War II (1939-44) and more recently, the expectation of life has been increasing and at the same time the infant mortality rate has also been quickly declining. Two vital indicators of economic development of a country are:

(i) Low Infant Mortality Rate

(ii) High Expectation of Life.

Hence from these two points of view we can say that we, the Indians, have attained some sort of development. This is, indeed, a matter of happiness to the aged persons. We may recall here that Rabindra Nath Tagore had once told, "I do not want to die in this beautiful world. I want to remain alive among the human beings".

It is a fact that technological advancement and improvement in health services has been reducing death rates among the elderly. Besides, there is a considerable change in the physical and sociological circumstances of the people with the transformation of traditional joint family system into nuclear families. The gradual marginalisation of the aged people in taking most of the decisions in an average family and the breakdown of the family as a traditional social unit that used to take care of the elderly has nowadays brought forth the problems of the elderly in the society. It is also appropriate to mention that with the rising number of old persons and their changing socio-economic and physical context, the proportion of the destitutes among them has been rising. In many families, such persons are anyhow being tolerated, not liked.

THE AGED PERSONS ARE OUR WEALTH—THEY SHOULD BE DULY HONOURED

In Bengali there is a proverb that states: "Take advice from those who have three heads." Indirectly, it suggests that we should seek advice from the very old persons who have gathered rich experience in various fields throughout their lives. Besides, according to the "Theory of learning by doing" introduced by Professor Arrow, another Nobel Laureate in Economics, if the retired persons are allowed to implement their wise experience in different spheres of our society, the society will itself be richly benefited. We get the instances of many retired persons who are freely willing to give consultancy services that should be duly adhered to. By recognizing such persons we ourselves will be productive and benefited.

Many researchers have shown that many small-scale industries are nipped in the bud not due to lack of capital but due to immature and inferior type of management. In this regard also by taking free advice of the retired persons who had profound knowledge in business, such concerns can thrive and flourish.

It has also been found from various research papers that during infancy and just before the time of sending the children to the educational institutions, the frequent accompaniment of the children with the highly experienced and wise persons may help to arouse so many good qualities in them. The wise counseling of such persons in many charitable organizations bringing up many children, may give rich dividend.

For the proper development of any country we are to consider not only the urban areas but also the rural areas where the proper advice of honest, lion-hearted and very active personalities are badly needed. The lack of such advice on the one hand and the unwillingness of the young persons to stay and work in the rural areas on the other, has not yielded proper development in the rural areas. In this respect also the proper advice of the aged persons may do a lot.

We have learnt from various sources that many countries of the world have been properly and adequately utilizing the rich experience of the aged people. The United States of America, England, New Zealand, Russia and Japan are pioneers in this direction. The Japanese Government have prescribed the rules for employing the aged persons and in New Zealand the mentioning of age in the advertisement for services has been forbidden. In England, many schemes have been implemented, where retired persons can get employment according to their intelligence and experience. In the USA discrimination in the field of service with regard to age has been made illegal and measures have been taken there to give service to "senior citizens".

A research paper published by the International Labor

Organization discloses that the so-called common concepts with regard to the old age are, to a large extent, full of errors. These organizations carrying intensive research work in Canada, France, Belgium, Germany, Netherlands, Sweden, Britain and the USA, have concluded that old age does not, normally, put obstruction to acquire learning. These organizations have also concluded that to increase expertise of the aged persons, special training is needed. In Russia, there is an institution called "The Soviet Institute of Gerontology" whose one of the research papers "The Secret of Long Active Life" is highly relevant here. From this paper it can clearly be known that the old age is not the age of seclusion and gloom, rather it is the time for service and constant self-development.

Many may opine that what we have discussed so long is highly relevant to the very rich countries. In spite of that we may say that in the relatively less-developed countries there is no point in thinking that the aged persons will have to lead a secluded life. Our life expectancy has been increasing and at the same time the number of unemployed persons has been increasing. Considering all such things we are to think of how we can best utilize those aged persons having strong physique and alert mind. In fine, we may state that the government of India has, of late, been recognizing the aged persons who have been duly treated as "senior citizens". To convey a deep sense of gratitude too, they are being paid a token advantage in some fields of our life. This is, indeed, a welcome step taken by our government. The younger persons and the children should also remember this and not do anything that will hurt such citizens. Again, it is in remembrance of the aged persons that 1999 had been honoured as the "International Year for the Aged Persons".

REFERENCES

1. Ackley, G. *Macro Economic Theory*, old edition.
2. *Census of India*, various issues.
3. Datt, R. and Sundharam, K. P. M. (2002): *Indian Economy*.

4. Datt, R. and Konar D. N. (2000): *Social Choice and Development,* Deep and Deep Publications Pvt. Ltd. New Delhi.
5. *Human Development Report*, Various Issues.
6. Konar, D. N. (1993): "Problem of Food and Malnutrition in the SAARC Countries", *Artha Beekshan*, Vol. 6, No. 2.
7. *Span Magazine*, Various Issues.
8. *World Development Report*, Various Issues.

Part B: Urbanization and Related Issues

5

URBANIZATION IN INDIA

INTRODUCTION

One of the most significant of all post-war demographic phenomena and the one that promises to loom even larger in the future is the rapid growth of cities in the developing countries. In 1950, 275 million people had been living in the cities of the third world countries, which was a mere 38 percent of the 724 million total urban population. According to the estimates of the United Nations, the world's urban population had reached 1.56 billion by 1975 and more than 50 percent of these people lived in metropolitan areas of the developing countries. Another estimate made by the United Nations reveals that in 2000 over 2.12 billion or 66 percent of the urban dwellers of the world will reside in less developed regions. This is the general trend of urbanization across the world. Needless to say, India and for that matter West Bengal is no exception. Obviously, India (and West Bengal) has been following the same track.

URBANIZATION: ITS MEANING, MEASUREMENT AND LIMITATION

The concept of urbanization in India is quite old. It started with the setting up of a Municipal Corporation in Chennai in 1688. In our censuses different criteria for urbanization have been accepted. Since 1971 the following criteria have been adopted for a place to be termed as urbanized:

(a) All places with a municipality, corporation, canton board or notified town or committee,

(b) All other places which satisfy the following criteria:

(i) A minimum population of 5000,

(ii) At least 75 percent of male working population engaged in non-agricultural pursuits, and

(iii) A density of population of at least 400 persons per square kilometer.

Demographically, urbanization means an increase in the proportion of the urban population in the total population of a country over a period of time. But this definition does not give out the whole truth of the process of urbanization either in India or in any of the states of India. This point has been well illustrated by Professor Ashish Bose.

The census figures of urbanization in India in 1961, 1971 and 1981 revealed that during 1961-71 and 1971-81 inter-censal periods the proportion of total population which was urban, increased from 18.0 percent to 19.9 percent and further to 23.3 percent. On the other hand, the proportion of the population residing in Class I towns having a population of 1,00,000 or more, increased from about 48 percent to approximately 66 percent and further to more than 60 percent. Bose has pointed out some limitations of using the said method as a measure of urbanization, specially, in India. This measure fails to indicate the intensity of urbanization because it does not bring to light the growth of big cities and the simultaneous stagnation of small towns. Also there is the fact that the census authorities have adopted various definitions of urban areas in different census years. This also puts a limitation to the study of a continuous process of urbanization in the country.

TREND OF URBANIZATION IN INDIA

Economic development is generally associated with the growth of urbanization. Some economists observe that the acid test of development lies in the shift of population from the rural areas to the urban areas. In the following paragraph the trend of urbanization in India and West Bengal in the twentieth century has been charted out.

During the first half of the twentieth century, India passed through a period of economic stagnation. Consequently, the extent of urbanization was very limited. Table 5.1 will demonstrate this phenomenon.

Table 5.1: Relative Growth of Urban and Rural Population in India

Years	Population in millions			Decade-wise increase in urban population (Percentage)	Urban Rural Ratio
	Total	Rural	Urban		
1901	232.9	207.3	25.6	-	1 : 8.1
1911	246.0	220.4	25.6	00	1 : 8.6
1921	244.3	216.6	27.7	+ 8.25	1 : 7.8
1931	270.8	237.8	33.0	+ 19.1	1 : 7.2
1941	309.0	265.5	43.5	+ 32.1	1 : 6.1
1951	361.1	298.7	62.4	+ 43.2	1 : 4.7
1961	439.2	360.3	78.9	+ 25.3	1 : 4.5
1971	548.2	439.1	109.1	+ 38.0	1 : 3.7
1981*	685.2	525.7	159.5	+ 46.8	1 : 3.3
1991**	846.3	628.7	217.6	+ 35.6	1 : 2.9
2001	1027.1	741.7	285.4	+31.2	1 : 2.6

* includes projected population of Assam.
** includes population projected for Jammu & Kashmir.
Source: Registrar General, India.

The above table reveals that urban population had increased from about 26 million in 1901 to about 62 million in 1951—an increase of only 36 million in 50 years! But, thereafter, the absolute increase during the next three decades (between 1951 and 1981) was of the order of 97 million. This may indicate that the programmes of industrialization did make an impact in terms of population absorption in urban areas, though this impact is not very discernible. In the next decade (between 1981 and 1991) alone, urban population increased by 58 million which indicates a growing trend towards urbanization. Urban population, in absolute terms, reached the figure of about 218 million accounting for 25.7 percent of total population. Again between 1991 and 2001 urban population increased by 68 million. In 2001 urban population accounted for about 28 percent of total population in India.

We may now analyse the process of urbanization in India through the percentage of people living in urban and rural areas. Table 5.2 will highlight it.

Table 5.2: Percentage of People Living in Rural and Urban Areas in India and West Bengal

Census years	India		West Bengal	
	Rural	Urban	Rural	Urban
1901	89.01	10.99	87.80	12.20
1911	89.59	10.41	86.95	13.05
1921	88.66	11.34	85.59	14.41
1931	87.81	12.19	84.68	15.32
1941	85.92	14.08	79.59	20.41
1951	82.72	17.28	76.12	23.88
1961	82.04	17.96	75.55	24.45
1971	80.10	19.90	75.25	24.75
1981	76.72	23.28	73.53	26.47
1991	74.29	25.71	72.52	27.43
2001	72.21	27.79	72.00	28.00

Source: For India computed from figures taken from first four columns of Table 5.1. For West Bengal, Census of India 2001, Series-20, West Bengal.

The table explains that during the first half of the twentieth century the extent of urbanization was very limited to India. The first post-independent census of India, which was taken in 1951, reveals that only 17.0 percent of the total population of India lived in urban areas. This percentage improved nominally and became about 18 percent in the following census. From then there has been a gradual improvement in this percentage. This picture is, however, different in West Bengal. Henceforth this percentage had started improving. The level of urban expansion in the State of West Bengal had kept well ahead of India during the first sixty years from 1901 to 1961. It is only more recently that the two rates have converged as has been revealed through Table 5.2.

DECADAL VARIATION OF THE GROWTH RATE OF URBAN POPULATION

The extent of urbanization in India and West Bengal can also be studied through the decade-wise annual growth rate of people living in urban areas. This has been presented in Table 5.3.

Table 5.3: Decade-Wise Variation of the Growth Rate of Urban Population (in Percentage)

Census Years	India	West Bengal
1941-51	2.85	3.53
1951-61	3.12	2.41
1961-71	2.54	3.31
1971-81	2.79	3.86
1981-91	2.62	3.17
1991-2001	2.75	1.87

Source: Computed from census data.

The above table clearly emphasizes that barring the decades 1951-61 and 1991-2001, the decadal annual compound growth rate of urban population of West Bengal has always exceeded that of India.

TOWNS CLASSIFIED IN INDIA AND THEIR PRESENT TREND

In Indian censuses towns have been classified into six classes on the basis of the number of people in a particular place. This information is portrayed in Table 5.4.

Table 5.4: Classification of Towns in India on the Basis of Population

Population	Nature of Town
1,00,000 and more	Class I
50,000 to 99,999	Class II
20,000 to 49,999	Class III
10,000 to 19,999	Class IV
5,000 to 9,999	Class V
Below 5,000	Class VI

Source: Census of India.

Significantly, in India, the total number of people of a particular place is the solitary criterion for giving it the required status of the class of town it belongs to.

We may now show the trend of various classes of towns we have been having in the five censuses after independence. Table 5.5 will exhibit this trend.

Table 5.5: Number of Towns in India*

Years	Class I	Class II	Class III	Class IV	Class V	Class VI	All Classes
1951	76	91	327	608	1124	569	2795
1961	102	129	437	719	711	172	2270
1971	148	173	558	827	623	147	2476
1981	216	270	738	1053	739	229	3245
1991	296	341	927	1135	725	185	3609

Source: Registrar General of India.
* The corresponding figures for 2001 census are not yet available.

We come to know from Table 5.5 that the importance of towns belonging to class V and VI has been dwindling with the passage of time while the importance of the first four classes of towns has been increasing. Class I towns have been the seats of administration and general economic activity. Naturally, industry, transport, trade and commerce, higher education (both general and technical), administrative and liberal services have been concentrated in them. This explains the increasing concentration of urban population in this size-class.

Let us now examine the percentage distribution of people of India in different classes of towns in each and every census conducted in the twentieth century. This has been demonstrated in Table 5.6.

A careful study of this table reveals that the percentage of people living in class I towns has been increasing. This is clear from the fact that in 1901 only 26 percent of urban people used to live in these towns whereas in 1991 this percentage had moved to 65.2—an increase of about 250 percent. Obviously, there is a trend towards high concentration of population in the bigger towns. The percentage of population in class II and III towns has remained almost stagnant while the importance of class IV towns has gradually been diminishing. Since 1951 the percentage of people living in class VI towns has been negligible and that living in class V towns has been slowly declining.

Table 5.6: Percentage Distribution of Population in Different Classes of Towns

Years	Nature of Towns						Total
	Class I	Class II	Class III	Class IV	Class V	Class VI	
1901	26.0	11.3	15.6	20.8	20.2	6.1	100
1911	27.5	10.5	16.4	19.7	19.3	6.6	100
1921	29.7	10.4	15.9	18.3	18.7	7.0	100
1931	31.2	11.6	16.8	18.0	17.2	5.2	100
1941	38.2	11.4	16.4	15.8	15.1	3.1	100
1951	44.6	10.0	15.7	13.6	13.6	3.1	100
1961	51.4	11.2	16.9	12.8	6.9	0.8	100
1971	57.3	10.9	16.0	10.9	4.4	0.5	100
1981	60.4	11.6	14.3	9.6	3.6	0.5	100
1991	65.2	10.9	13.2	7.8	2.6	0.3	100

Source: Computed from Census data published in various issues.

Let us now see whether the importance of different classes of towns in our country has been declining with the passage of time. To do that, ranks have been given to each of the six types of towns in the census years of the twentieth century. Table 5.7 will highlight this phenomenon.

A glance at Table 5.7 shows that the position of class I and class VI towns has remained unaltered in each census year discussed here. Class III towns had occupied the fourth rank in the first four censuses but in the remaining censuses such towns had found higher rank, that is, rank two. Class II towns had occupied the fifth rank in the first six censuses but in the subsequent censuses such towns had improved their rank.

Table 5.7: Ranks Attained by Towns of India in the Census Years

Years	Nature of Town					
	1st	2nd	3rd	4th	5th	6th
1901	1	5	4	2	3	6
1911	1	5	4	2	3	6
1921	1	5	4	3	2	6
1931	1	5	4	2	3	6
1941	1	5	2	3	4	6
1951	1	5	2	3	4	6
1961	1	4	2	3	5	6
1971	1	4	2	3	5	6
1981	1	3	2	4	5	6
1991	1	3	2	4	5	6

Source: Ranks computed from census data published in various issues.

We may now discuss total population that has been living in different classes of towns in the census years of the post-independent India. Table 5.8 will display the scenario.

Table 5.8: Urban Population in India (lakhs)

Years	Nature of Towns						All classes
	Class I	Class II	Class III	Class IV	Class V	Class VI	
1951	275	61	97	84	80	19	616
1961	399	87	132	99	53	6	776
1971	612	117	171	117	48	5	1070
1981	945	182	224	149	56	8	1564
1991	1388	233	281	165	55	6	2128

Source: Census of India, Various Issues.

An analysis of Tables 5.5, 5.6 and 5.8 reveals that the towns at the margin in class II have been jumping into class I towns. This is evidenced by the fact that, whereas there were only 76 towns of class I type in 1951, their number had grown to 296 in 1991—an increase of nearly three times. As a consequence, the total population of class I towns had increased from 275 lakhs in 1951 to 1388 lakhs in 1991—an increase of 405 percent.

Class II and class III towns are those in a transitory stage. Their number and also their total population have gone up in absolute terms. The total number of class II towns had increased from 91 in 1951 to 341 in 1991 and the total population from about 61 lakhs to 233 lakhs—a four-fold increase. The number of class III towns had increased from 327 in 1951 to 927 in 1991 while their total population had increased from 97 lakhs in 1951 to 281 lakhs in 1991—an increase of 190 percent.

The proportion of urban population of class IV, V and VI towns has shown a downward trend. The number of class IV towns had increased from 608 in 1951 to 1135 in 1991 while their total population had increased from 84 lakhs to 165 lakhs—an increase of only 96 percent. As against it, the number of class V towns had declined from 1124 in 1951 to 724 in 1991 whereas their total population had declined from 80 lakhs to 55 lakhs, during the period under consideration—a decline of about 30 percent. Ironically, the number of towns of class VI type had declined from 569 in 1951 to nearly 185 in 1991. During the same period the number of total population of class VI towns had suffered a sharp decline from 19 lakhs to only 6 lakhs!

TREND GROWTH RATE OF POPULATION OF DIFFERENT CLASSES OF TOWNS BETWEEN 1901 AND 1991

We may now explain the trend growth rate of urban population for a long period of ninety years starting from 1901 and ending at 1991. For the sake of analysis the entire period has been divided into two parts: (1) Between 1901 and 1941 and (2) Between 1951 and 1991. Besides, there is the overall period of ninety years. The various growth rates have been put in Table 5.9.

An analysis of Table 5.9 displays that between 1901 and 1991 the growth rate of urban population in India had been 2.8 percent compounded annually. The corresponding figures for six types of towns had respectively been 4.4, 2.9, 2.6, 1.5, 0.0 and -1.3 percent. In comparison to the trend of urbanization in the pre-independent

India, the trend in the post-independent India is quite distinct. In the latter period the trend growth rate of urban population was 3.7 percent whereas in the former period this was only 1.4 percent. However, in the post-independent period, the last two types of towns ceased to be popular.

Table 5.9: Growth Rates of Population in Different Classes of Towns in India

Nature of Towns	Growth Rates and Periods		
	Period I: 1901 to 1941	Period II: 1951 to 1991	Overall Period: 1901 to 1991
Class I	2.5	5.7	4.4
Class II	1.6	4.1	2.9
Class III	1.5	3.1	2.6
Class IV	0.7	1.9	1.5
Class V	0.6	-0.7	0.0
Class VI	0.0	-1.8	-1.3
All Classes	1.4	3.7	2.8

Source: Computed from census data.

INCREASE IN NUMBER OF LARGE CITIES IN THE POST-INDEPENDENT INDIA

One significant characteristic of urbanization in the post-independent India has been the increasing trend of the larger cities having a population of one million plus. In 1951 there were only five such cities, namely (1) Kolkata, (2) Mumbai, (3) Chennai, (4) Delhi and (5) Hyderabad. In 1961 two more cities, namely Bangalore and Ahmedabad were added to the list of five already in existence. In 1971, Kanpur and Pune were also included to the list of five. Again, in 1981, the total number of such cities had gone up to eleven with the inclusion of Nagpur and Lucknow. Finally, in 1991, the total number of such cities had become twentythree, the twelve new cities being Surat, Jaipur, Kochi, Coimbatore, Vadodra, Indore, Patna, Madurai, Bhopal, Vishakhapatnam, Varanasi and Ludhiana.

In Table 5.10 we present the population of twelve large cities of India in 1981 and 1991.

Table 5.10: Population of Twelve Large Cities of India in 1981 and 1991

(Unit: Million)

Cities	1981	1991
Greater Mumbai	8.2	12.6
Kolkata	9.2	10.9
Delhi	5.7	8.4
Chennai	4.3	5.4
Hyderabad	2.9	4.3
Bangalore	2.5	4.1
Ahmedabad	2.5	3.3
Pune	1.7	2.5
Kanpur	1.7	2.1
Nagpur	1.3	1.7
Lucknow	1.0	1.6
Jaipur	1.0	1.5

Source: Census of India, Different Issues.

INDIAN CITIES AMONG THE CITIES OF THE WORLD

We are familiar with the population map of the world where India has the second place, next only to China. However, two of our great cities, namely, Greater Mumbai and Greater Kolkata had come in the list of twelve largest cities of the world, having a population of ten million or more. Table 5.11 authenticates this picture.

We see from Table 5.11 that in the list of twelve most densely populated cities of the world, Greater Mumbai has got the sixth place while Greater Kolkata has got the tenth place. Naturally, these two Indian cities have become overcrowded with population and they have been suffering from numerous problems, some of which are being analyzed in the following paragraph.

Table 5.11: Population of Twelve Large Cities of the World in 1991

(Unit: Million)

Cities	Population
Mexico	20.2
Tokyo	18.1
Sao Plao	17.4
New York	16.2
Sanghai	13.4
Greater Mumbai	12.6
Los Angeles	11.9
Buenos Airs	11.5
Seoul	11.0
Greater Kolkata	10.9
Beijing	10.8
Rio de Zenerio	10.7

Source: The World Development Report, Various Issues.

EFFECT OF URBANIZATION

Regarding the effect of urbanization professor M. Todaro has pointed out "Along with the rapid spread of urbanization has come the prolific growth of huge slums and shanty towns. From the favellas of Rio de Janeiro and the Pueblos jovenes of Lima to the bustees of Kolkata and the bidonvilles of Dakar, such makeshift communities have been doubling in size every five-ten years. Today slum dwellers represent over one third of the urban population in all developing countries; in many cases they account for more than 60 percent of the urban total."

This is exactly true to the urban areas of West Bengal, especially to Kolkata about which we shall have a detailed discussion in the following chapter.

DEGREE OF URBANIZATION IN ORISSA AND BIHAR

We shall now dwell on the extent of urbanization in two of the Eastern States of India. This has been done on the basis of census data on four classes of towns as revealed through censuses of 1951, 1961, 1971 and 1981. These two States are Bihar and Orissa. Here we have shown that urbanization can be measured by the famous Pareto Law and that the slope of the Pareto Curve can be indicative of the degree of urbanization by which we mean the rapidity with which the cities and towns in a country have been thriving and also have been attracting people.

PARETO CURVE IN THE CONTEXT OF MEASUREMENT OF URBANIZATION IN INDIA

Before we apply the important Pareto Curve for measuring urbanization in India, it is better to say something about this curve and its fitting.

PARETO CURVE: ITS FITTING AND INTERPRETATION

The Pareto Law introduced by V. Pareto in 1897, is usually associated with the distribution of income in an economy and it shows that there is a distinct inverse relationship between the level of income (after a certain limit) and the corresponding number of income-receivers. According to Pareto, the distribution of income in a stable economy is given, approximately, by the formula

$$N_x = \alpha x^{-\beta} \qquad (1)$$

for sufficiently large values of x which is income, N_x is the number of persons earning x or more, α and β are two constants.

The form (1) can also be applied to generate various kinds of cities and towns according to different population levels. This means that the cumulative frequency distribution of towns by size class of population actually responds to the Pareto Law. In this form the number of towns at different levels of population has replaced the number of income-receivers and the level of population in the corresponding towns has substituted the level of income. Consequently, a distinct and coherent relationship between them has been obtained by the simple application of the Pareto Law, which is, in general, operative in the case of towns with a population of 10,000 and above.

Let T stand for the number of towns having population greater than or equal to P so that the expression (1) becomes

$$T = \alpha P^{-\beta} \qquad (2)$$

Taking log on both sides we get

$$\log T = \log \alpha - \beta \log P \qquad (3)$$

The values of α and β can be determined by applying the famous least squares technique. The two normal equations from which the values of the parameters α and β can be obtained are:

$$\Sigma \log T = n \log \alpha - \beta \, \Sigma \log P \qquad (4)$$

$$\Sigma(\log T \log P) = \log \alpha \, \Sigma \log P - \beta \, \Sigma(\log P)^2 \qquad (5)$$

From (3), we note that,

$$(d \log T) / (d \log P) = -\beta \qquad (6)$$

This means that the slope of the Pareto Curve measuring in log values of towns vertically and log values of population horizontally is $-\beta$. The negative sign of the slope indicates an inverse relationship between the number of towns and the population. Again since

$$| (d \log T) / (d \log P) | = | dT / T | / | dP / P | = \beta,$$

the absolute value of β may also be thought of as the elasticity of number of towns with respect to population variation. The degree of urbanization can thus be gauged by the absolute slope of the Pareto line.The relative number of towns in different classes varies from one census to another. Accordingly, we get different values of β in different censuses and different Pareto Curves as well. With the passage of time the different lines may not resemble each other: one may be steeper than the other, or one may be flatter than the other or one may even be parallel to another. The steeper the line (that is, the greater the absolute value of β) the more even will be the distribution of the urban population in the first four classes of towns. On the other hand, the flatter the line (that is, the lower the absolute value of β =) the more uneven will be the distribution of urban population in the first four classes of towns and there will be a high concentration of population in some particular class (classes) of town.

APPLICATION OF PARETO LAW IN ESTIMATING DIFFERENT TYPES OF TOWNS

We have applied the Pareto Law in form (2) to estimate the number of towns in class I containing a population of 1,00,000 or more, in class II where the number of population lies between 50,000 and 99,999, in class III giving abodes to a population varying between 20,000 and 49,999 and class IV with a population lying between 10,000 and 19,999 in two relatively poor states of India as mentioned earlier, namely Orissa and Bihar. In case of both the states we have fitted the Pareto Curve for each census data between 1951 to 1981 and we have seen that in both cases the Pareto Curve has been a very good fit. Our fitting of the Pareto curves has been done on the basis of the classification of the first four classes of towns as had been done in different censuses. The necessary data for this in both the states have been taken from various censuses and the different issues of Statistical Abstracts of India. We have chosen the census of 1951 as the starting point of our discussion because in this year it was noticed that there had been an unprecedented growth of towns and cities in India and an unprecedented growth of urban population too. This may be illustrated by the fact that whereas, during the 1941-51 inter-censal period the total population of India increased by 13.3%, the urban population increased by 41.4%.

The results of the Pareto Curve fitted on the distribution of population according to four classes of towns in the states of Orissa and Bihar have been portrayed in Tables 5.12 and 5.13 respectively.

A glance at Table 5.12 reveals that for Orissa the absolute value of β increased in regard to the two successive censuses of 1951 and 1961. Consequently, the Pareto Line for 1961 had been steeper than that for 1951. The percentage structure of urban population in 1961 displays that this population was more or less evenly distributed in four classes of towns—the maximum concentration of urban population was found in class IV towns and the minimum in class I towns.

Table 5.12: Actual and Estimated Towns in Orissa

Year	1951		1961			1971			1981		
Class	Actual	Estimated	Actual	Urban Population (%)	Estimated	Actual	Urban Population (%)	Estimated	Actual	Urban Population (%)	Estimated
I	1	1	1	13.19	1	5	38.3	4	6	41.63	6
II	1	1	3	20.55	2	1	3.9	4	7	12.76	7
III	5	5	8	20.27	10	19	29.2	14	23	21.82	22
IV	8	8	22	28.00	22	21	15.5	24	39	17.03	41
Total	15	15	22	82.01	35	46	86.9	46	75	93.24	76

Source: Computed from actual figures taken from various censuses.

Table 5.13: Actual and Estimated Towns in Bihar

Year	1951		1961			1971			1981		
Class	Actual	Estimated	Actual	Urban Population (%)	Estimated	Actual	Urban Population (%)	Estimated	Actual	Urban Population (%)	Estimated
I	5	5	9	43.5	9	11	45.40	11	16	54.12	17
II	6	6	7	12.9	8	9	11.05	11	19	14.38	17
III	19	19	28	21.6	26	42	23.95	36	57	19.26	49
IV	37	36	42	14.9	43	57	14.54	63	59	10.04	81
Total	67	66	86	92.9	86	119	94.94	121	151	97.80	164

Source: Computed from actual figures taken from various censuses.

The distribution of urban population in class II and class III towns was of identical nature. The absolute slope of the Pareto Line in 1971 had abruptly fallen from 1.486 to 1.045 with the consequence that the urban population got concentrated principally in class I and class III towns and the percentage of population had radically fallen in class II and class IV towns. The absolute value of β in 1981 had slightly improved over that in 1971. Such an improvement meant a relatively even distribution of urban population. The Table clearly shows that the percentage of urban population in class I and IV towns had improved marginally whereas that in class II towns had risen significantly and that in class III towns had fallen rapidly. Hence so far as Orissa is concerned, we see that distribution of urban population had high concentration around class I and class III towns only in 1971 when the absolute value of β was the least.

Coming to Bihar, we notice that β had attained the least absolute value only in 1981 which also displays high concentration of urban population in class I towns as more than 54 percent of urban population had lived in this class of towns. The distribution of urban population in 1961 and 1971 censuses was more or less of a uniform pattern. In each case class I towns had topped the list in the ranking of the distribution of urban population in the four types of towns. The class III towns had taken the second place while class IV and class II towns had occupied the third and fourth ranks respectively.

If drawn with log T presented vertically and log P horizontally, the Pareto lines for each census and for each state will be seen to have the shape of a straight line. We may, therefore, infer that the number of different classes of towns in these two states of Eastern India can well be generated by the Pareto Law.

A COMPARATIVE STUDY OF THE PROCESS OF URBANIZATION IN ORISSA AND BIHAR

A comparative study of the process of urbanization in these two states can now be made.

In Orissa the absolute values of β in regard to each of the four censuses have been greater than one. But in the case of Bihar this value was greater than one only in regard to the 1961 and 1971 censuses while in regard to the 1951 and 1981 censuses this value was less than one. Such values of β indicate that the distribution of urban population in the State of Orissa was more even than Bihar.

In Bihar in regard to each of the three censuses considered here the urban population was concentrated, mainly, in class I and class III towns. However, in 1981, the percentage of urban population in the class III towns fell rapidly and that in Class I towns increased abruptly. In contrast, Orissa saw in 1961, a high concentration of urban population only in class IV towns while class I towns had the least percentage of urban population. In terms of the two subsequent censuses, however, the percentage of urban population in Orissa like that in Bihar was concentrated, first, in class I towns and then in class III towns. Again, in both these states in 1971 the class IV and class II towns had respectively the third and fourth rank with respect to the distribution of urban population. Orissa, in 1981, still maintained the previous rank but in Bihar in terms of this census these two types of city simply interchanged their ranks.

REFERENCES

1. Bhattacharyya, Basabi (1991): "Urbanization and Human Development", *Economic and Political Weekly*, Vol. XXXIII. Nos. 47 and 48.
2. Bose, A. 1974: *Studies on Indian Urbanization, 1901-1981*, McGraw Hill, New Delhi.
3. *Census of India*: Various Issues.
4. Chatterjee, Mahalaya (1996): "Urbanization and Demographic Changes in India (1961-1991)", *Business Studies*, Calcutta University, Vol. XIX, No. 1 & 2.
5. *Economic Review*, Government of West Bengal, Various Issues.
6. Giri, Pabitra (1998): "Urbanization in West Bengal, 1951-1991", *Economic and Political Weekly*, Vol. XXXIII, Nos. 47 & 48.
7. Goon, A. M. & Halder, D. (1990): "Calcutta's Population: Changing Size

and Age Sex Composition", *Arthashastra*, Vol. 9, No. 2.

8. *Human Development in South Asia*, 1997.
9. Konar, D. N. (1991): "Degree of Urbanization in Orissa and Bihar–an Econometric Analysis", *Arthashastra*, Vol. 10, Nos. 1 & 2.
10. Konar, D. N. (1996): "Urbanization in West Bengal and the Problem of Environment", *Economic Affairs*, Vol. 41, No. 1, January-March.
11. Kosambi, Meera (1994): Urbanization and Urban Development in India, Indian Council of Social Science Research, New Delhi.
12. Todaro, M. (1987): *Economic Development in the Third World*, Orient Longman.
13. *World Development Report*, Various Issues.

6

POPULATION PRESSURE AND THE PROBLEM OF ENVIRONMENT IN KOLKATA

INTRODUCTION

Calcutta (Kolkata since 1st January 2001), the capital city of West Bengal and one of the oldest cities of the world, has now-a-days, been suffering from an environmental crisis. The basic reason behind this is the fact that she can no longer bear the burden of staggering population. This city was originally built for a population of about one million in order to give her status of a metropolitan city. But our Kolkata had crossed that mark way back in 1911 when her population was 1,016,445. It is to be noted that Kolkata was one of the eleven metropolises the world already had 100 years ago in 1900, together with such modern giants as New York, Berlin, Chicago, Tokyo, London and Paris. Today (according to 1991 census) Kolkata (Calcutta Metropolitan Area, CMA) with a population of 10,916,272 has become the tenth largest urban agglomeration of the world (the first being the Mexico city, the Greater Mumbai being the sixth one). This has been indicated in the previous chapter. The core Kolkata has, however, a population of about 3.3 million. The tremendous pressure of population has been eating into the vitality of Kolkata and has been making her environmentally degraded and polluted.

TREND OF POPULATION GROWTH IN CALCUTTA

The trend of the growth of population in Kolkata from 1872 (when the first census had been taken in India) to 2001, the time of the last census, has been portrayed in Table 6.1.

Table 6.1: Total Population and the Annual Compound Growth Rate of Population in Percentage (Decade-wise) in Kolkata

Years	Total Population	Growth Rate
1872	704750	-
1881	684658	-0.29
1891	741144	0.80
1901	933754	2.34
1911	1016445	0.85
1921	1053334	0.36
1931	1221210	1.47
1941	2167485	5.92
1951	2698494	2.22
1961	2927289	0.82
1971	3148746	0.73
1981	3305006	0.49
1991	4399819*	0.00 (2.90)
2001	4580544*	0.40

* Total CMC.
Source: Calculated from various censuses.

The earlier estimate of the population of Kolkata is found in 1716 when 12,000 people used to live here. Then in less than 100 years (in 1801) the population of this city increased to 140,000. In 1821 the population of this city had grown to 179,917 and in 1850 there lived 500,000 persons in this city.

The growth of population in Kolkata from 1872 to 2001 and the corresponding annual compound growth rate, in percentage, from 1872 to 1881 and at each decade from 1881 to 2001 can easily be understood from Table 6.1. The population of 1991 and 2001 presented in the above Table represent figures of total Calcutta Municipal Corporation (CMC) having an area of 185 sq. km while

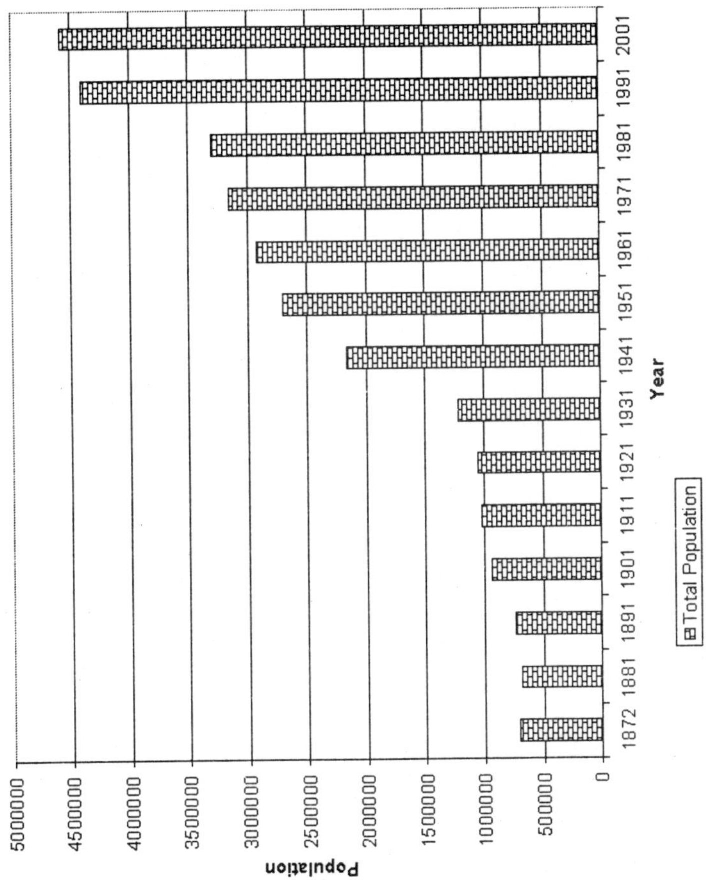

Diagram 6.1: Vertical Bar Diagram Representing Total Population of Kolkata in Various Census Years.

those of the earlier censuses are the figures of core CMC with an area of only 104 sq. km. The Calcutta Urban Agglomeration (CUA) has an area of 852 sq. km. and within it 10,916,272 people live, as reported earlier. In the two consecutive censuses of 1981 and 1991 population living in core Calcutta had remained almost identical. To emphasize the growth of population in core Kolkata within an area of 104 sq. km. over the last 100 years from 1891 to 1991, we have used the vertical bar diagram measuring year horizontally and the absolute number of population (in million) vertically. This has been demonstrated in Diagram 6.1. The height of the bars themselves signify the gravity of the situation. It is a relief that since 1961 the growth rate of population in core Calcutta has started falling and between 1981 and 1991 this rate has become zero. The annual growth rate of population, 0.40 percent, between 1991 and 2001 represents the growth rate of population of CMC with an area of 185 sq. km.

PAVEMENT DWELLERS OF KOLKATA

A significant part of population of Kolkata is those who live on pavements. According to the latest census of 2001, more than 7.5 lakh people of Kolkata live on its pavements. Such people are in Kolkata with nowhere to go and no one to help them. This scene compared on a country-wide scale will be even more horrific because Kolkata is only fourth in the National Census of pavement dwellers in cities. Surat tops the list with 11.80 lakh, Jaipur is second with 11.50 lakh and Mumbai ranks third with a population of 11 lakh. The number of people and number of pavement dwellers in some important metropolitan cities of India can be known from Table 6.2.

How alarming the situation is! How do these destitute maintain their livelihood? It is not possible to respond to such questions so easily. But the fact is that Kolkata is overburdened with the pressure of population.

REASONS BEHIND EXPLOSIVE POPULATION IN KOLKATA

The growth of urban population is usually attributed by three factors, namely (a) natural growth of urban population, (b) migration from other states and (c) migration from rural to urban areas within the state itself. However, behind the rapidity of the growth of population in Kolkata lies the basic fact that Kolkata acts as a job magnet at the national level. Besides, it is the best Center of Trade and Administration. Naturally, it has attracted job aspirants both from rural areas and from the neighbouring states. That is why job seekers like bureaucrats, office employees, military men, transport workers, mechanics, plumbers, domestic servants, middlemen, etc. have thronged here for years together and have still been thronging here.

Table 6.2: Total Population and Total Pavement Dwellers in Different Metropolitan Cities of India in 2001 (unit: number)

Metros	Population	Pavement Dwellers
Surat	2811000	1180000
Jaipur	2324000	1150000
Mumbai	16368000	1100000
Kolkata	4580544	750000
Chennai	6425000	500000
Hyderabad	5334000	350000
Bangalore	5687000	300000
Ahmedabad	4519000	275000
Kanpur	2690000	225000
Pune	3756000	195000

Source: Census, 2001 (Published in The Statesman, Kolkata, March 18, 2003).

Besides, living in the countryside has become more difficult for thousands of people who constitute the rural labor force. The small and medium-sized towns of Kolkata's hinterland are hardly in a better position to offer employment to millions of rural people. The growth of Kolkata's population is thus only partially due to the influx from outside. It is surprising to note that in this core city with an area of 104 sq. km. had lived a population of 1.02 million in 1911 and in this very city with the same area lived a population of 3.3 million both in 1981 and 1991. In this context it will be worthwhile to mention that since our villages are underdeveloped, people are forced to come to Kolkata during the day time to earn their livelihood and the total population during the day rises to about 6.5 million in core Kolkata having the same space of 104 sq. km. The city's daytime population is about 10 percent of West Bengal's total population of 681 million (1991 census). Many of these people swarm the pavement and kerbs throughout the day.

It is a fact that even in these days of soaring prices Kolkata has become a city almost equally preferable to all sorts of people, specially to the poor for whose needs a fairly diversified but low-priced commerce is in operation in all parts of the city. In Kolkata, one can exist even by begging. Naturally, numberless people come to this city mainly from 24-Parganas and Midnapore in West Bengal, from Ganjam and Balasore in Orissa, from Muzaffarpur, Bhagalpur and Darbhanga in Bihar and Ballia in Uttar Pradesh, almost in tattered clothes and with empty pockets. There are, in this city, about forty-five thousand handcarts and about thirty-five thousand rickshaws which are manned by most of these people. Wayside tea stalls, cigarette and betel shops and eating places run with their labor. Domestic service, cheap and back-breaking, depends basically upon their toil and trouble. It is they who provide the city's sweepers, washermen and most of the milkmen. Kolkata is the only city in the world, where within 104 sq. km. there are four wards with a population of 175,000 and twentyfive wards with more than 80,000 people per sq. km. respectively. Such being the nature of Kolkata's in-migrants; it is but natural that Kolkata is to suffer a lot from the high pressure of population.

SOCIAL STATUS OF CORE KOLKATA AND KMA POPULATION

Let us now bring out the social status of the people of Kolkata. To do that we should, first of all, display the economic condition of the people in Kolkata Metropolitan Area (KMA) as presented in Table 6.3.

The modal income of the people, obviously, falls in the income range of Rs.150 - Rs.350. This being the economic condition of most of the households of the KMA, what they can do with this meager income or what they can do to supplement their income is anybody's guess. 50 percent of the people of core Kolkata are to live in slums in sub-human conditions, 70 percent of such people live below the poverty line and, on an average, 10 people sleep in a room of [10′ × 10′]. In the decade between 1981 and 1991 the annual growth rate of the people of core Kolkata was almost zero but that for the people living in the slums was 9.3 percent. In KMA, 63 percent of the children below the age of six are malnourished and in this area 66 percent of the families live in a one-roomed unit. In core Kolkata, 56,000 people sleep on the footpath and live on pavements in 3′ ×3′ ×3′ huts made of plastic sheets and bamboo poles. Naturally, they use road-side lanes as their toilets.

Table 6.3: Household Income of KMA People per Month Based on 1984 Data

Income Range (Rs.)	Percentage of Households
150-350	45
350-600	26
600-1500	24
1500 and above	5
Total	100

Source: Arranged from information published in the Statesman, June 9, 1996.

ENVIRONMENTAL PROBLEMS ASSOCIATED WITH POPULATION PRESSURE

The tremendous pressure of population in this city has deteriorated her environment in every possible manner. We have already pointed out that 50 percent of the people of core Kolkata have been living in slums which are already overcrowded. In these slums we find innumerable chullas burning low-grade coal, wood, bamboo, wastes of materials, etc. In many of these slums people rear cows, buffaloes, goats, sheep and so on. All things naturally make the neighbouring areas dirty, smoky, noisy and smelly. The surrounding roads also become impassable. The people living in the slums are absolutely apathetic about the quality of life in their surroundings because their most important concern is how to fill their stomach and thus they have no time to think of anything else.

Side by side the life of the slum-dwellers, runs the life of persons belonging to the upper middle class and that of the richer ones. To make their lives easy and comfortable, they ply innumerable vehicles of different makes along the narrow streets of the city. With the passage of time the number of vehicles running along the roads and streets of the city has been growing at an increasing rate but the space of these roads and streets has remained unchanged. We may mention here that Kolkata has the lowest ratio of road area to city area—it is less than 6 percent. While plying on the streets and roads of the city, the ever-growing army of motor vehicles disgorge somewhere between 650 and 700 tonnes of pollutants into the air, making it troublesome to breathe.

Let us turn to the city's daily traffic as shown in Table 6.4.

Table 6.4 Daily Traffic on Roads of Kolkata at a Glance

Nature of Vehicles	Number
Buses	8444
Minibuses	1200
Taxis	27633
Luxury Taxis	7000
Trucks	41500
Motor Cycles	365000
Vans	25500
Autos: Licensed	15000
Non-licensed	20000
Total Road length of Kolkata	1,404km
Ideal Road length	22% of city area
In reality	5% of city area

Source: The Ananda Bazar Patrika (a Bengali daily), dated 17-3-2003.

The content of this table is self-explanatory. Together with this the drivers of buses, minibuses and autos while plying their vehicles on the road very often try to overtake their counterparts. Such a practice not only creates panick among co-passengers but also leads to serious accidents. They forget the statistical law of average and do not even learn by their presious experience. They are simply incorrigible.

Again, a big source of city's pollution is more than 200 tonnes of garbage that is dumped on the city's streets every day. People, almost of all walks of life of the city, have been accumulating garbages here and there every day and soon they turn into garbage-hills on which many persons unhesitatingly go on committing nuisance, which even oozes in the nearby road. These garbages and the accumulated nuisance remain uncleared for days together. Naturally, the areas surrounding these garbage-hills become adversely affected by the bad smell exuding from there.

Moreover, the high density of population of Kolkata has gradually swallowed up the open space where public can take some recreation. Kolkata has only 6 percent of open space compared to 14 percent in Mumbai and 25 percent in Delhi. This is the fundamental reason why toxicity in Kolkata's air is on the increase. Besides, thousands of people maintain their livelihood here by establishing small and big factories. Smoke arising from these factories supplements the pollution to the air. Wastes of most of these factories get accumulated in the nearby garbage-hill and sound created by these factories adds to the sound pollution generally created by loud horns blown by lorries, cars, buses and other vehicles plying very often discretionarily along the roads.

Because of all the above reasons the air surrounding Kolkata is very much polluted. We may mention here that according to the World Health Organization (WHO), out of the seven cities of the world with the worst air pollution Kolkata (and Delhi) is one. We may also add that according to the United Nation's Report brought out at the end of January, 1994, Kolkata (also Mumbai and Delhi) was among the most polluted cities in the world.

It is sad to state that Kolkata's environmental problem had been felt as early as 1935 when in a report of Kolkata Corporation it was mentioned that to save Kolkata, "Its roads need to be widened, no more buildings to be constructed within Kolkata, more greenaries the city needs". The pity is that even after almost 57 years of Independence we have not realized the serious consequence towards which the city has been heading. The saddest thing to mention is that every now and then attempts are made to acquire parks and other open spaces for commercial construction and even in the twenty first century there is no symptom of stoppage of such actions.

REMEDIAL MEASURES

There can be no denying the fact that environment of our dear city has deteriorated considerably. Our immediate task is, therefore, to save her by hook or by crook. For saving her from further degradation we propose the following:

1. The excess burden of population from this city should, in no time, be released by properly building up satellite townships surrounding Kolkata. These towns should have good roads to connect themselves. Besides, a good number of offices, business center, different types of institutions should be shifted to these towns so that the pressure of population on Kolkata may be reduced to a great extent. We may point out there that the new township at Salt Lake has taken a great share of population of core Kolkata.
2. The living condition of the villages across Bengal should no longer be neglected. Some essential items, e.g. drinking water, primary education, medical and health services and transport facilities should be made available even in the remotest parts of Bengal.
3. Sufficient job opportunities in rural areas should also be created.
4. More and more emphasis should be put on the development of agriculture, rural industry and rural development.
5. People should be made more cautious about their health. They should also be made aware of their civic sense. Environment-damaging activities should forthwith be banned and in this respect help of laws and acts should be taken and necessary amendments to rules, if any, can also

be made. The violation of these rules should be heavily penalized.

6. In fine, we may state that "live and let live" should be our attitude and by adhering to this principle we should do everything good for the well-being of this city and, of course, for the well-being of ourselves.

REFERENCES

1. Ghosh, Murani, Dutta, Alok. K and Ray Biswanath (1972): *Calcutta - A Study in Urban Growth Dynamics*: Firma K.L. Mukhopadhyay, Kolkata.
2. Goon, A.M. and Halder, D. (1990): "Calcutta's Population: Changing Size and Age Sex Composition, *Arthashastra*, Vol. 1 No. 2.
3. Konar, D.N. (1988): "Population Pressure and the Problem of Environment in Clacutta", *Journal of Commerce*, Vidyasagar University, Vol. 3 March.
4. Munshi, S.K. (1975): *Calcutta Metropolitan Explosion*, Peoples' Publishing House.
5. Ray, Nisith Ranjan (1988): *Calcutta—The People of a City*, K.P. Bagchi and Company, Kolkata.

7

WOMEN AND THE RURAL TRANSFORMATION IN INDIA

INTRODUCTION

Women constitute an integral part in the socio-economic life of any country in the world. However, their role in economic development was not seriously considered and "Women's place is home", was a common and popularly accepted slogan during the period of Industrial Revolution. According to Ester Boserup, this view has been due to the patriarchal set-up of such societies through the system of land rights. Women and development as a subject entered the international scene around 1970 and from then on the 'hidden' economic and social contribution of women began to be uncovered from under-layers of social mythology. The declaration of the "International Women's Year" by the United Nations' General Assembly in 1975 could be considered a turning point regarding the place, role and position of women in a society. Since then it has come to be recognized that half of the humanity of every country was not so long connected with development. They are the harbingers of human culture and active partners of economic development. Even this notion is gaining ground that "development without women" cannot take place.

WOMEN IN INDIAN HISTORY

A thorough study of the Indian History from the ancient time reveals that women in our country have played a very significant role in various periods. Lord Buddha was very much pleased with the milk-made rice served by Sujata under a tree; sisiter Rajarshi was always a spontaneous source of inspiration to King Harsha; Sultana Rijiya was a great heroine; Lakshmibai sacrificed her life for saving India during the time of the Sepoy Mutiny; Matangini Hazra was a famous patriot and forsook her life with the tri-coloured National Flag in her hand during the period of the Freedom Movement; the kindness of Bhagawati Devi, the mother of Pandit Ishwar Chandra Vidyasagar, a great son of Bengal, is known to everybody; Sarada Devi is still now memorable and worshipped. There is a long list and these are merely for an illustration. Besides, in our epics, namely the *Ramayana* and the *Mahabharata*, we get the names of many illustrious ladies who are, even today, whole-heartedly worshipped by most of the citizens of our country.

In this context we may state that Indian women in 1930 had given a spontaneous response to Mahatma Gandhi's call for the Civil Disobedience Movement. Being surprised at this participation Panditji observed, "I think that the most important and significant feature of that movement was the tremendous part that the women in India took in it. It was astounding." Nehru subsequently stated his views on the status of Indian women in a speech which he had given in 1949: "I am quite convinced that in India today progress can be and should be measured by the progress of the women in India."

BASIS OF THE INDIAN ECONOMY AND THE PRESENT TREND

India is a land basically dependent upon villages and it is said that India lives in its villages. Even today about 72 percent of the total population of India lives in villages where agriculture and its allied activities are the principal sources of livelihood. Again 25 percent of our total population belong to the Scheduled Castes and Scheduled Tribes, most of whom also live in villages alone. At present there is, however, a continuous spree for living in urban areas and this has led to rapid urbanization in our country as has been reported in Chapter 5.

WOMEN HAVE NOT BEEN DULY ATTENDED TO

In spite of the strong potentiality that the women have in our country, we have failed to utilize that. We have come to know that out of 49 crores of women in our country about 16 crores live in abject poverty. Moreover, about 70 percent of our women are illiterate. It is really regrettable that about 50 percent of pregnant women in our country suffer from anaemia. We have already seen that the sex ratio has gradually been going unfavourable to women. Again, the statistics of the work participation rate (WPR) indicate that this rate for women in 1991, became 14.8 percent in comparison to 11.8 percent in 1981. Moreover, the percentage distribution of women as non-workers by type of activity in the rural-urban residence in 1981, indicates that in comparison to 22.94 percent of urban females as students only 10.7 percent of rural females fall in this category. Also 27.66 percentage of urban females are dependants, infants and disabled while 32.07 percent of rural ones fall in this group. All these facts reveal that we have given less emphasis on the progress and development of women in our country. This amounts to saying that about 50 percent of the population of the world's second most populous country has not been properly utilized and we have shown continuous negligence for them. This is particularly true to the women living in rural areas.

To recognize the role of women in economic development and to enhance the status of women all over the world four International Conferences on Women have so far taken place in addition to the numerous sessions on women and development in various International Conferences. The first conference took root in Mexico in 1975, the second one was held in Copenhagen in 1980, the third one took place in Nairobi in 1985 and the fourth one in September 1995 in Beijing. Besides, in the World Summit that took place in Copenhagen on "Poverty and Unemployment", the role of women for the economic development of any country was duly highlighted.

We have already seen that in India there has been always an explosive growth of population which has been eating into the fruits of development. Two probable reasons behind such growth of population in India (especially after 1951) are:

a) The norm of a small family has not yet reached to many.

b) The high fertility rate of women, especially in rural areas.

WOMAN IS THE SAVIOUR

To keep the population of our country within a reasonable limit, naturally the growth rate of it should be checked. The women, in general, and those living in the rural areas, in particular, can, through the dissemination of education, check the growth of population to a great extent in our country. Let us now see how this can come true.

INFANT (CHILD) MORTALITY IN INDIA

One possible reason behind the non-acceptance of the norm of small family by many in our country is the prevalence of high infant mortality rate which means that an infant will die within the first year of its life. The symbol $q^{(x)}$ representing the estimated value of the probability of a new-born child dying before reaching the age of x multiplied by 1000 is very significant in connection with the infant (child) mortality rate. $q^{(1)}$ is thus closer to the infant mortality rate. Naturally, in case of infant and child mortality, the values $q^{(1)}$, $q^{(2)}$, $q^{(3)}$ and $q^{(5)}$ representing, respectively the number of deaths per 1000 live births by ages 1, 2, 3 and 5 are very crucial.

The nature of infant (child) mortality in India in 1980 can be guessed from Table 7.1.

Table 7.1: Estimates of Child Mortality in India in 1980

Infant Mortality	Urban			Rural			All Areas		
	P	M	F	P	M	F	P	M	F
$q^{(1)}$	67	68	66	123	130	116	115	122	108
$q^{(2)}$	80	83	77	135	138	133	123	125	120
$q^{(3)}$	84	85	84	147	145	149	132	130	134
$q^{(5)}$	98	97	99	167	161	173	152	147	157

P = Person, M = Male and F = Female.

Source: Census of India, 1981 (Occasional Papers No. 5 of 1988).

It appears from Table 7.1 that the child mortality in India is not a stable entity. It varies from areas to areas and also from sex to sex.

Again, there are wide variations in the values of $q^{(1)}$, $q^{(2)}$, $q^{(3)}$ and $q^{(5)}$ among the states and Union Territories in India. Among the states in India, Manipur has attained the minimum values of q which are 32, 40, 44 and 50 for all persons at the respective age of the child concerned. Next to Manipur is Kerala where these values are 52, 55, 63 and 80 respectively. On the other hand, Madhya Pradesh, Uttar Pradesh and Arunachal Pradesh have attained very high values of child mortality.

LEVEL OF EDUCATION OF MOTHER AND CHILD MORTALITY IN INDIA

Child mortality depends upon a number of complex and interrelated set of biological, demographic, social, economic and cultural factors. We shall, however, try to explore that it is the level of education of mothers that helps reduce the child mortality in our country.

Child mortality has been seen to be an inverse function of the level of education of the mother. This means that with the increase in the education level of the mother there is a sharp fall in child mortality. This will be evident from Table 7.2.

Table 7.2: Estimates of $q^{(2)}$ and $q^{(5)}$ by the Level of Education of the Mother in India in 1980

Child Mortality	Educational Level of Mothers				
	Illiterate	Literate but below middle	Middle but below matric	Matric but below graduate	Graduate and above
$q^{(2)}$	132	99	63	43	28
$q^{(5)}$	170	107	71	48	32

Source: Ibid.

It follows from Table 7.2 that child mortality is about five times higher among illiterate mothers as compared to mothers having educational level upto graduation and above. Thus the level of education of the mother has a very strong impact on lessening the child mortality. One fundamental reason behind lower child mortality in urban India is that mothers in this area are far more literate than those in rural areas. Various studies conducted by different scholars in many less developed countries reveal that maternal education plays a very significant role in determining infant and child mortality.

Moreover, in most cases educated mothers themselves rarely hesitate about taking contraceptive devices in order to lower the size of families and thereby they succeed not only in lowering the size of their families but also in reducing the mortality of their children.

LEVEL OF EDUCATION OF WOMEN AND BIRTH RATE

Educational level itself has got a direct role to play in reducing birth rates. It is a fact that education affects fertility by influencing the desire for having small-sized families and at the same time by increasing the age of marriage.

The direct effect of age in reducing fertility can be guessed from the normal behaviour of fertility data which, for all countries, show that usually age-specific fertility begins from a low point, rises to a maximum somewhere in the age-group 20-29 last birth-day and thereafter declines steadily. Table 7.3 will make this point clear.

Table 7.3: Fertility Rates Specific for Age of Mother in India in 1980

Age Group	15-19	20-24	25-29	30-34	35-38	40-44	45-49
Age specific fertility per 1000 females	88.2	246.1	227.6	163.1	97.1	44.3	20.2

Source: Registrar General's Newsletter. Vol. 15, No.2 (April 1984).

It appears from Table 7.3 that the larger the age at marriage the lower will be the birth-rate. It has been calculated that an increase in the mean age at marriage of females by 3 years over the period from 1891-1901 to 1951-61 in India had resulted in a 3 to 4 percent reduction in birth rates over the same period. Table 7.4 illuminates the effect of age at marriage and non-marriage on the reduction of fertility in India.

Table 7.4: Estimated Total Marital Fertility Rates and Total Fertility Rates per Woman in Rural India in 1978

Level of Education	Fertility Rate (Total Marital)	Total
Illiterate	5.5	4.7
Literate	5.0	3.9
Primary	4.9	3.6
Matric and Above	4.7	2.5
All women	5.4	4.6

Source: India, Registrar General, 1982.

The information contained in Table 7.4 brings out both the direct and the indirect role of education (by enhancing the age at marriage and keeping many women unmarried) in reducing fertility.

CONCLUSION

It follows from what has been pointed out in the preceding paragraphs that education leads to a sharp fall in the child mortality and thereby assures the parents of their children's survival. This assurance, in most cases, influences the parents to adopt the norm of a small family. Moreover, education leads to an enhancement of the age at marriage and thereby causes a reduction in the fertility of women. Thus a quick fall in birth rate can be attained through the dissemination of education, especially among women. Therefore, in the present set-up of the Indian economy it is the duty of all concerned to see that at least primary education is spread to women of all walks of life. This will, it is hoped, save us from the present trend of explosive growth of population in our country by lowering the birth rate to a good extent.

Moreover, it has been observed that in most of the families (particularly in the poor and poorer ones), almost the hundred percentage of income earned by the women through strenuous efforts, is spent for the welfare of the family itself, neither for their own enjoyment nor for their own luxury. Besides, women in a family of target group play a vital role in augmenting income of the family. It has been noticed that in the poor family income of the women has a positive correlation with nutrition, status of the family, educational status of the children. All these create a soothing atmosphere in the family.

The mother is the dearest person to a child, particularly in its infancy and the child imbibes the ideas of its mother. So the mothers through their love and affection can make their children hard-working, honest, sincere and obedient — some rare distinctions which most children of today lack. If the children attain these qualities through the continuous influence of their mothers, not only will the rural India be improved but also the whole of India will be prosperous and happy. A healthy child can be born to a healthy mother. Hence, if the mothers suffer from fear, agony, depression and anaemia, they cannot give birth to healthy children. So let us try our best to make all mothers (and for that matter all women) healthy, educated, faithful, respectful and true partners of our life. The women should also come forward with a constructive attitude and it is through these supportive activities based on wide co-operation, mutual understanding and mutual respect that will help us in transforming the present India into a much-desired one.

It is heartening to know that of late women's role has been widely recognized in our country and also in different parts of the world. We have been celebrating the eighth day of March every year as the International Women's Day with much fun-fare. It is worth mentioning that in a path-breaking step the Gujarat Govt. has made it compulsory to mention one's mother's name during the time of admitting children in schools across Gujarat. Besides, in many educational institutions for filling in application forms, the applicants are to mention the names of both father and mother. Moreover, to show reverence to women of our country our Finance Ministers at the center have been giving special tax concessions to the in-service women. In the budget for 2003-04, our former Finance Minister Mr. Yashwant Singh has announced that upto

Rs. 12,000 per annum spent on child's education will be deducted from taxable income for couples with up to two children. This has a number of implications. In fine, we may add that Amartya Sen, the Nobel Laureate Professor has, for the last twenty years or so, been laying special stress on women especially of the third world countries. According to him, as long as there will be deprivation of women and bias towards the male children, growth is bound to suffer. He has repeatedly emphasized that empowerment of women will remain a dream until the girl child is given equal opportunity in education and treated on equal terms in relation to the male child. While delivering the 7th Justice Sunanda Bhandare Memorial Lecture in New Delhi on "Consequences of Gender Inequality", Professor Sen emphasized, "Lack of education among girls is directly linked to gender inequality. Female literay has helped in reducing the gap between men and women." He added, "There is an urgency in rectifying the inequality, enhancing their power and status in society, which can not be achieved by involving the women agencies alone but with common efforts by all." He stressed that the role of women in economic growth could not be confined to household activities, as they have to play a major part in the economic growth outside the family.

REFERENCES

1. Agarwala, S. N. (1977): *India's Population Problems* (2nd Edition), Tata Mcgraw-Hill Publishing Co. Ltd., New Delhi.
2. Boserup, Ester (1970): *Women's Role in Economic Development*, George Allen & Unwin.
3. Caldwell, John C, Reddy, P. H. and Caldwell, P. (1982): The Causes of Demographic Change in Rural South India- A Micro Approach, *Population and Development Review*, 8.
4. *Census of India*, Various Issues.
5. Datta, Amlan (1995): "Social Order"- I and II, *The Statesman*, Kolkata, September 18,19.
6. Dutt, Mohendra Nath (1982): *Society and Education I: Society and Woman II*, The Mohendra Publishing Committee.

7. Government of India, Office of the Registrar General (1982): Levels, Trend and Differentials in Fertility, 1979, Ministry of Home Affairs, New Delhi.

8. Konar, D. N. (1989): "Role of Education in Reducing the Growth of Population in India", *Arthashastra*, Vol. 8, No. 2.

9. Sen, Amartya (2002): "Consequences of Gender Inequality", the 7th Justice Sunanda Bhandare Memorial Lecture, *The Statesman*, Nov. 15.

8

UNEMPLOYMENT PROBLEM IN INDIA AND DEVELOPMENT OF THE TOURISM INDUSTRY

I. EMPLOYMENT SCENARIO OF OUR COUNTRY

One very crucial factor hitting our entire economy is the gradual increase in the magnitude of unemployment. In this context the Sixth Five-Year Plan stated: "The pace of generation of employment opportunities seems to have lagged behind the growth of labour force... The number of people unemployed and underemployed has risen significantly over the last decade."

Productive employment is the only sustainable means in any country. Naturally, employment generation has always been one of the top priorities of our Planning Commission. All the political parties pledge before the holding of every election to eradicate unemployment, if voted to power. But ironically, the reality is that the number of the jobless in our country has been increasing since 1991. In some cases "golden handshake" has been given to many. The educated persons finding no job opportunities, has become a liability to the society. The economic reforms with the LPG package introduced in July 1991 with much fanfare have failed to relieve the situation. In fact, the recent survey reveals that it has only aggravated the problem as it has resulted in an unfair competition between the strong and the weak, and the latter being overrun by the former has been lagging far behind. According to the projection of unemployment for 1990-2000, 28 million had been assumed to be the backlog of unemployed in 1990 and net additions to the labour force during 1990-95 had been expected to be 37 million. Thus the total number of persons requiring employment during the Eighth Plan period (1992-97) would be around 65 million. It is expected that during 1995-2000, labour force would increase by 41 million. Hence by the year 2000, the total number of job seekers would be around 106 million. The Planning Commission, thus, concludes, "Employment Growth in the aggregate over the estimated employment of 300 million would have to be about 4 percent compound per annual if the goal of providing employment to all is to be achieved by the end of the Eighth Plan and over 3 percent per annum if it is to be attained by 2000 A.D." (Ref: Planning Commission: Employment, Past Trends and Perspectives for 1990's). The Union Labour Minister has, of late, expressed serious concern over the widening gap between the number of job seekers and the availability of job opportunities in India.

There can be no denying the fact that the fundamental reason behind the increasing magnitude of unemployment is the very high rate of growth of population, which has, in the last fifty two years, increased at an annual compound rate of 2.38%. A natural consequence of such unemployment is the widespread poverty prevailing throughout our country as has been demonstrated in Table 3.5 (vide chapter 3).

(i) Organised Sector and Employment Generation

It is a fact that the bulk of employment in our country is provided by the organized sector, comprising the Central Government, the State Government, Quasi-Government bodies, local bodies and the private sector. But this sector itself has got some obvious limitations to provide employment to all who are actually fit for getting the same. The magnitude of employment in this sector over the last few years has been shown in Table 8.1.

It is seen from Table 8.1 that employment in the public sector has remained almost stagnant as only about two crores of the labor force of our country are engaged in this sector. The annual compound growth rate of employment in this sector between 1990 and 2000 has been found to be 0.57 percent. The picture is slightly better in the private sector where the number of people employed had increased from nearly 76 lakh in 1990 to nearly 87 lakh in 1998 and remained almost the same in 1999. However, in the following year (that is, in 2000) there was a marginal fall in the number. For the entire period of eleven years between 1990 and 2000, the annual compound growth rate of employment in the private sector has been calculated to be 1.61 percent. However, the annual trend growth rate of employment in the organized sector as a whole has just been 0.64%.

Table 8.1: Employment in Organized Sector (million)

Year	Public Sector	Private Sector	Total
1990	18.77	7.58	26.35
1991	19.06	7.68	26.74
1992	19.21	7.85	27.06
1993	19.33	7.85	27.18
1994	19.45	7.93	27.38
1995	19.47	8.06	27.53
1996	19.43	8.51	27.53
1997	19.56	8.69	28.25
1998	19.42	8.75	28.17
1999	19.42	8.70	28.12
2000	19.31	8.65	27.96

Source: Employment Review, DGE & T, Ministry of Labour, Government of India, New Delhi.

(ii) Gravity of Unemployment Situation of Our Country

It is undoubtedly true that the unemployment problem has crippled our country. In the preceding section we have, already, indicated the gravity of the situation. Let us now try to magnify it to some extent by taking an example of the condition of the qualified engineering students. Of late, there has been a craze among the students passing the pre-degree course for making themselves engineers and so on. It is because of this that in many states the number of engineering colleges has been gradually increasing. For example, in West Bengal the number of such colleges has, of late, increased from four to thirty six. In this way the supply of degree and diploma holding engineers has, obviously, been increasing. But is there sufficient demand for them? The picture is not so rosy as it appears to be. Table 8.2 will bring this to light.

Table 8.2: Unemployed Engineers in Different States of India in 1995

States	Degree holders	Diploma holders	Total
Karnataka	3427	8882	12309
Andhra Pradesh	2743	6737	9480
Uttar Pradesh	904	7093	7997
Tamil Nadu	2294	4809	7103
Maharashtra	2120	4895	7015
Gujarat	1599	4270	5869
Kerala	1061	3692	4753
Madhya Pradesh	1094	3211	4305
Bihar	915	3829	3744
West Bengal	273	3045	3318

Source: Manpower Profile of India, 1997.

Table 8.2 illustrates the plight of the unemployed Degree and Diploma holding engineers in 1995 in various states of our country. Karnataka heads the list of such engineers while the position of West Bengal is at the bottom. It is to be noted that the number of Diploma-holding unemployed engineers is about three times that of the Degree-holders in major states, except for Uttar Pradesh where the former is about eight times that of the latter. If this be the lot of the qualified engineers, that of other educated persons or non-educated persons can be easily imagined.

(iii)The Problem Of Child And Bond Labor

The problem of child labor in our country has been discussed, in detail, in Chapter 3. Here we shall, briefly, touch upon the case of bonded labourers. It is a matter of great shame to state that even at the beginning of the twenty first century bonded labour is also in existence in our country. A survey conducted by the Gandhi

Peace Foundation and National Labour Institute states that there are about 26.7 lakh bonded labourers in different states surveyed by them. However, records of the Ministry of Labor, Govt. of India, indicate that only 22,000 bonded labourers had been identified at the end of February 1988. Thus that the magnitude of such labourers is quite high cannot be doubted though there is difference in the methods followed in exactly quantifying it.

II. TOURISM INDUSTRY AND ITS ROLE

What has been analyzed in the previous chapters amounts to saying that the very failure of checking the growth of population in the right moment, when we had preached the slogan of bringing family planning lest the Malthusian catastrophe befalls us (and even today), has given rise to such cumulative effect in our society that our very existence has been put to danger. The economist's high sounding talk like "trickle down effect" has not given rich dividend, nor have the poverty alleviation schemes yielded the magical effect it were expected to do. We should remember that charity cannot bring sustainable relief to anybody. It is the employment, and productive employment alone, that can bring sustainable relief to the people gaining employment. In this sphere the tourism industry can, obviously, play a very crucial role.

The tourism industry has, now-a-days, become one of the most significant industries across the world. Universally, its place is just below the oil and automobile industries. To maintain national integration and to preserve and enrich national understanding the importance of this industry can never be exaggerated. To enhance national income as well as employment and to earn invaluable foreign exchange the role of this industry cannot be underestimated. It is an extremely labor-intensive industry and its linkage effect is highly pervasive. According to some prediction made by the World Tourism Organization (WTO), tourism in the year 2020 will be an industry netting five billion dollars a day. This industry in the 21st century will not only be the world's biggest industry, it will also be the largest. This industry will have a major impact on the economic, environmental, sociological and cultural aspects of the society, which will be increasingly scrutinized by the Government, consumer groups, environmentalists, media persons and the travelling public. The economies of many a country of the world, e.g., Hong Kong, Singapore and Sri Lanka, are dependent very much on this industry. Besides, the economic conditions of many small islands, viz., Mauritius, Fiji and Malta, are largely sustained directly and indirectly by this industry alone. It may be mentioned in this context that in Mauritius there is a travelling garden whose area is 720 square kilometers. As an individual industry in India the place of this industry is only after gem and jewellery and ready-made garments. Moreover, this industry tops the list of the industries bringing foreign exchange into our country.

According to a report from the Ministry of Tourism, Govt. of India, the number of people already employed in the tourism industry in India, exceeds seven million accounting for about 2.4% of our total labor force. Around eleven million people of our country are directly and indirectly dependent upon this industry. The significant feature of this industry is that it employs a large number of women, both educated and uneducated, skilled and unskilled. Women, in fact, outnumber men in hotels, airline services, travel agencies, handicraft and cultural activities.

It has been estimated that the employment multiplier figure in the tourism industry sector is 2.36. This means that a direct employment of 100 persons in the tourism industry creates jobs for 136 persons in other sectors of our economy. This is, indeed, a great gain.

There is a common saying that the Bengalees are fond of touring and travelling at the opportune moment. This is true especially in these days when the joint family system has broken down and there is the general tendency to have a small family (called nuclear family) consisting of the parents and one or two children. Many families will set out even for a short trip at the weekend. Some like to dine at a hotel or a restaurant on the Sundays and holidays. People, instead of meeting their friends and relatives occasionally, prefer to roam hither and thither just for forgetting the drudgery of their working life and also for getting some relaxation and recreation. This has become the habit of the people of the lower middle class of our society. For the upper middle class, starred hotels, restaurants and the like are welcome. They frequently go abroad for getting entertainment and recreation. For them air-conditioned trains, luxury coaches, first class compartments, airlines, etc., are necessary. For giving adequate comfort and happiness to all sorts of tourists thousands and thousands of manpower are required. Naturally, a number of industries are linked to this industry and quite obviously in such a flourishing and busy industry money gets transacted with extreme velocity which is actually needed for sustaining and generating income and employment. With the tourism industry are linked the hotel and restaurant industries, roads, transport and communication, fruits, flowers and vegetables, local products, medicines, books, magazines and so on. Tourists are prospective buyers of the local fairs. Even the so-called child labourers instead of doing the most strenuous and hazardous jobs, can earn a handsome amount by smilingly serving the tourists.

III. TOURISM INDUSTRY IN INDIA

With the establishment of a Department called "Tourism and Civil Aviation Department" more than 37 years ago (in 1967), our government recognized Tourism as an industry. India with her vast area, with her physical size and appearance, heritage, cultural, splendid natural resources, coastlines, minerals, forest cover, old and antique buildings, the Tajmahal—one of the seven wonders of the world, monuments, beautiful temples and mosques, forts, hill stations, scenic beauty, springs, rivers, canals, ponds, dams, flora and fauna, gardens, historical places, wild life sanctuaries can, easily, find a permanent place in the firmament of the tourism industry. From time immemorial many foreigners have been attracted by the beauty and natural resources of our country. By doing so they have themselves mixed with us. All this have given rise to the "Great India", of which we are really proud. But the tragedy is that people coming from different parts of the globe have plundered our Mother India and they have made themselves wealthy and rich. A sample of such loot can still be found in the beautiful palace of Cintra, which is very close to Lisbon, the capital city of Portugal. Here many of the items plundered by Vasco-da-Gama, the most clever and expert among the Portuguese navigators, at the closing years (1497) of the fifteenth century, have been well preserved. It is reported that on coming back to Portugal Vasco-da-Gama lamented and said, "I am ashamed of my own wealth. Practically, I have nothing in comparison to what India has." Dadabhai Naoroji has, beautifully, described the nature of such exploitation, loot and plunders in his famous book "Poverty and Un-British Rule in India". Even at present about one crore eighty lakh people of Bangladesh have been freely living with us. We should not brood over the past; it is better to forget that. We should make our India and us too as much attractive and fascinating to the tourists, especially to the foreigners, as possible. We should remember that we are in a fix as to what to do with the staggering population. We should try our level best for a healthy development of this industry.

In spite of the country having all potentiality to attract foreign tourists, the tourist traffic here has been less than what is expected. This aspect can be highlightened in Table 8.3.

Table 8.3: India's Share of Earnings in World Tourism (billion US dollars)

Year	World's earnings	India's earnings	Percentage share of India
1980	102.4	1.40	1.37
1985	116.1	1.20	1.03
1990	265.1	1.51	0.57
1993	305.8	2.00	0.65
1994	321.5	2.31	0.72
1995	372.0	2.75	0.74
1996	404.0	2.96	0.73
1997	443.8	3.00	0.68
1998	439.4	2.94	0.67
1999	454.7	3.00	0.66
2000	475.8	3.16	0.66

Source: *Statistical Outline of India*, 1999-2000, 2001-02 and 2002-03, Tata Services Limited, Department of Economics and Statistics.

The above table reveals that we only share about 0.72% of the global earnings from tourism. This figure remained almost static over the three years from 1994 to 1997 in spite of the fact that our earnings from tourism industry have gradually been increasing from 2.00 billion in 1993 to 3.00 billion dollars in 1997 and further to 3.00 billion dollars in 1999. It is a pity that only 0.4% of the world's tourists visit our country. It is to be noted that notwithstanding such a meager figure of the arrival of the foreign tourists in our country, about 80% of the new employment created each and every year comes through this industry alone. Just think of the situation that we could have attained (and still we can attain) if the percentage figure of the world's tourists visiting India can be increased. It is worth mentioning here that with the approval of tourists' visa in India a sum of about Rs. 25,000/- in foreign currency is practically accumulated in the Government treasury.

Table 8.4: India's Share in World Tourism

Year	Tourist arrivals (million*)		
	World	India	% Share of India
1980	284.8	0.80	0.28
1990	463.6	1.71	0.37
1993	512.9	1.76	0.34
1994	528.4	1.89	0.36
1995	567.0	2.12	0.37
1996	594.8	2.29	0.39
1997	612.8	2.37	0.39
1998	635.1	2.36	0.37
1999	668.5	2.48	0.37
2000	701.9	2.64	0.38

Source: *Statistical Outline of India*, 1999-2000, 2001-02 and 2002-03, Tata Services Limited.

*Tourist arrivals in India include those from Bangladesh and Pakistan from 1985 onwards.

In 1968-69, that is, in the second year of the recognition of tourism as an industry in our country we earned only Rs.26 crores as foreign exchange. This figure had risen to Rs. 9,186 crores and

Rs.10,000 crores respectively in 1995-96 and 1996-97. Between 1980-81 and 1988-89 the annual compound growth rate of earnings of foreign exchange from this industry was calculated to be 8.70%. With a serious effort this rate can be enhanced and that will really be desirable.

It is also to be noted that the number of foreign tourists visiting India in 1951 was 16,823. However, in 1996-97 about two crore twenty-eight lakh foreign tourists visited us. It is reported that in December 1995 alone about 2,51,000 foreign tourists visited India and in this month alone about Rs.1,074 crores of foreign exchange were earned. In fact, in 1995-96 we received 2.1 million foreign tourists. Besides, over 100 million domestic tourists visit different parts of our country every year.

Tourists all over the world like visiting India and this will be seen from Table 8.5.

Table 8.5: Number of Tourists Arriving in India from Various Countries

Countries	1994			1995			% change
	Number	%	Rank	Number	%	Rank	
U.K.	300.7	30.6	1	334.8	30.0	1	+11.3
U.S.A.	176.5	18.0	2	203.3	18.2	2	+15.2
Sri Lanka	89.0	9.0	4	114.2	10.2	3	+28.3
Germany	95.4	9.7	3	89.0	8.0	4	-6.7
France	73.1	7.4	5	82.3	7.4	5	+12.6
Japan	63.4	6.5	6	76.0	6.8	6	+19.9
Canada	56.4	5.7	7	63.8	5.7	7	+13.1
Italy	43.5	4.4	9	53.0	4.8	8	+21.8
Malaysia	40.8	4.2	10	50.0	4.5	9	+22.6
Singapore	44.2	4.5	8	48.6	4.4	10	+9.9
Total	983.0	100.0	-	1115.0	100.0	-	+13.4

Source: Statistical Outline of India, 1996-97.

Table 8.5 portrays the pattern of foreign tourists arriving in India from the major countries of the world. We notice here that in comparison to 1994 there was a steady increase in the number of tourists reaching India in 1995 from every country barring Germany for which there was a real fall in the number. In both the years the tourists from the United Kingdom and the USA had occupied the first two ranks respectively. In 1994 the third rank went to Germany, while in the following year, this place was taken by Sri Lanka. Similarly, in between 1994 and 1995, there was some marginal change in the ranks among the last three countries. The Spearman's rank correlation coefficient of the countries from which majority of the tourists reached India in 1994 and 1995 has been calculated to be 0.95, indicating that the pattern of the foreign tourists visiting India in the said two years has remained almost unaltered. This also follows from the percentage distribution of tourists visiting India in these two years.

Between 1994 and 1995 there has been, on the whole, a 13.4% increase in the arrival of foreign tourists in our country from the above-named ten countries of the world. The largest such increment had emerged from Sri Lanka, one of the SAARC countries and the smallest one, a negative one, did from Germany. It is undoubtedly true that if infrastructural facilities can be made conducive to the foreigners, we are sure to invite many such tourists and this will be helpful to us from many an angle.

REFERENCES:

1. *Economic Review*, Govt. of West Bengal, Various Issues.
2. *Economic Survey*, Govt. of India, Various Issues.
3. Naoroji, Dadabhai: *Poverty and Un-British Rule in India.*
4. *Statistical Outline of India*, Tata Services Limited, Various Issues.

Part C: Regional, Inter-regional and International Aspects of Growth of Population

9

GROWTH OF POPULATION IN WEST BENGAL

DEMOGRAPHIC SCENARIO OF WEST BENGAL

Trend Growth Rate Of Population Of West Bengal

After making a brief discussion about the growth of population in India specially, since Independence, we may now turn toward the demographic scenario of West Bengal. According to the census 2001, West Bengal is the most densely populated state of India. This state possessing only about 2.77 percentage of land area of India has to nourish about 7.80 percent of total population of India. We shall now explain the trend of growth of population of West Bengal during the last 100 years spanning between 1901 and 2001, which has been portrayed in Table 9.1.

The fifth column of Table 9.1 will indicate the cause of concern for the high growth rate of population of West Bengal. The annual growth rate of population had, in the decade between 1951 and 1961, been 2.88%, though in the subsequent decades the growth rates had been slower. In the first half of the last century the annual trend growth rate of population of West Bengal had been 0.91% while that between 1951 and 2001 had been 2.49%. The population growth curve in West Bengal in these fifty years has been found to be

$$P_t = 27.48\ (1.249)^{t-1951}, \text{ unit being million.}$$

Table 9.1: Growth of Population in West Bengal (Number)

Year	Persons	Male	Female	Decade-wise percentage growth rate of population
1	2	3	4	5
1901	16940088	8708978	8231110	-
1911	17998769	9349419	8649350	0.61
1921	17474348	9173148	8301200	-0.30
1931	18897036	9997035	8900001	0.79
1941	23229552	12545269	10684283	2.09
1951	26299980	14105519	12194461	1.25
1961	34926279	18599144	16327135	2.88
1971	44312011	23435987	20876024	2.41
1981	54580647	28560901	26019746	2.11
1991	68077965	35510633	32567332	2.33
2001	80221171	41487694	38733477	1.65

Source: Census of India, Various Issues.

If this trend rate of growth is maintained, the projected population of West Bengal in 2011 will exceed ten crores as it will be evident from Table 9.2.

The nature of growth of male and female population in West Bengal in the last eleven censuses starting from 1901 and ending at 2001 can be visualiased from the compound bar chart as revealed through Diagram, 9.1. The gap between the height of two consecutive bars indicates the declining trend of sex ratio in the West Bengal as will be reflected in the following chapter.

The Growth Rate of Male and Female Population in West Bengal

The composition of total population of West Bengal during the last eleven censuses has already been presented in Table 9.1. We now intend to show the decade-wise annual rate of growth of male and female population of West Bengal, which have been put in Table 9.3.

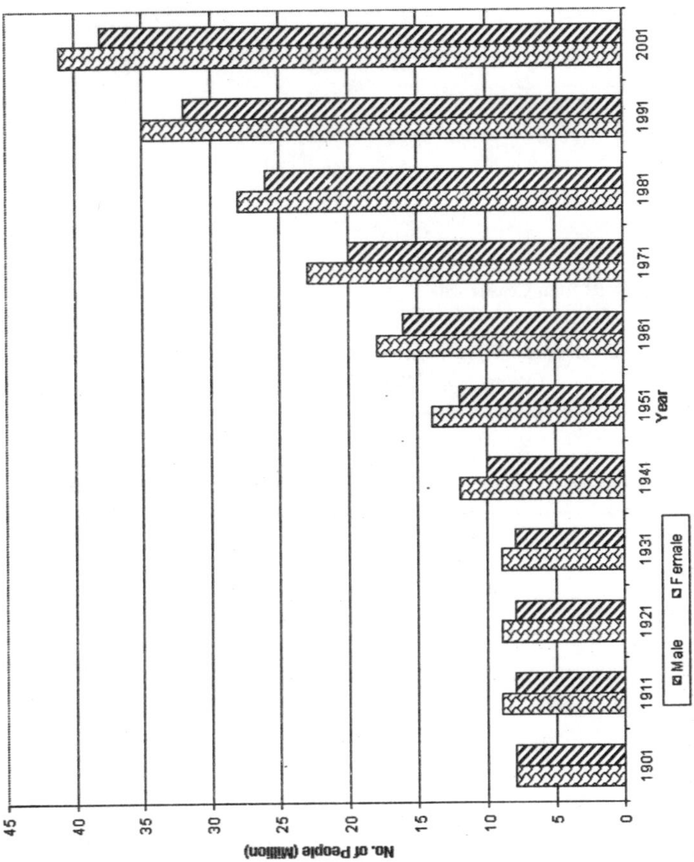

Diagram 9.1: Sex-wise populaiton of West Bengal

Table 9.2: Actual and Estimated Population of West Bengal Between 1951 and 2011
(unit: million)

Census year	Population	
	Actual	Estimated
1951	26.3	27.5
1961	34.9	34.3
1971	44.3	42.9
1981	54.6	53.6
1991	68.1	66.9
2001	80.2	83.6
2011	-	104.4

Source: Estimated from Census data.

Table 9.3: Decade-Wise Annual Rate of Growth of Population of West Bengal

Decade	Annual Growth Rate	
	Male	Female
1901-1911	0.71	0.50
1911-1921	-0.19	-0.41
1921-1931	0.86	0.70
1931-1941	2.30	1.84
1941-1951	1.18	1.33
1951-1961	2.80	2.96
1961-1971	2.34	2.49
1971-1981	2.00	2.23
1981-1991	2.20	2.27
1991-2001	1.57	1.75

Source: Computed from information contained in Table 9.1.

Table 9.3 reveals that in the first four decades the annual growth rate of male population was greater than that of female

Table 9.4(a): The Decade-Wise Growth of Population in the Districts of West Bengal Between 1901 and 1951

Districts	1901-1911	1911-1921	1921-1931	1931-1941	1941-1951
Darjeeling	5.31	5.12	12.85	17.72	17.58
Jalpaiguri	21.30	4.93	6.47	14.42	8.13
Cooch Behar	4.58	-0.07	-0.26	8.43	4.74
North Dinajpur	6.78	-12.13	7.23	11.92	17.03
South Dinajpur	6.78	-12.13	7.23	11.92	17.03
Maldah	15.72	-1.77	4.99	17.19	11.05
Murshidabad	1.71	-8.99	11.96	19.69	4.61
Birbhum	3.67	-9.41	11.25	10.63	1.77
Burdwan	0.37	-6.46	9.82	19.99	15.92
Nadia	0.36	-8.26	1.48	16.42	36.15
North 24-Parganas	14.78	6.31	9.53	26.07	23.50
South 24-Parganas	13.43	6.54	8.82	24.01	21.77
Hooghly	3.67	-0.90	3.22	24.02	13.35
Bankura	1.99	-10.43	9.00	16.00	2.30
Purulia	13.70	-5.98	16.91	11.95	7.43
Midnapore	1.15	-5.48	4.97	13.99	5.28
Howrah	10.93	4.71	10.17	35.62	8.12
Kolkata	10.71	3.63	11.30	24.30	25.68
West Bengal	6.25	-2.91	8.14	22.93	13.22

Source: Census of India, Various Issues.

Table 9.4(b): Decade-Wise Growth Rate of Population of West Bengal at the District Level Between 1951 and 2001

Districts	1951-1961	1961-1971	1971-1981	1981-1991	1991-2001
Darjeeling	35.90	25.16	31.02	26.91	23.54
Jalpaiguri	48.27	28.76	26.55	24.44	21.52
Cooch Behar	52.45	38.67	25.28	22.55	14.15
North Dinajpur	83.81	43.02	30.20	34.00	28.72
South Dinajpur	25.73	37.09	28.04	24.39	22.00
Maldah	30.33	31.98	26.00	29.78	24.77
Murshidabad	33.46	28.57	25.49	28.20	23.70
Birbhum	35.55	22.80	18.01	21.94	17.88
Burdwan	40.65	27.04	23.47	25.13	14.26
Nadia	49.81	29.91	33.29	29.95	19.25
North 24-Parganas	47.98	34.52	31.42	31.69	22.64
South 24-Parganas	30.75	34.24	19.77	30.24	20.89
Hooghly	39.02	28.72	23.86	22.43	15.72
Bankura	26.17	22.02	26.93	18.12	13.79
Purulia	16.33	17.86	15.65	20.00	13.96
Midnapore	29.26	26.89	22.39	23.57	15.68
Howrah	26.51	18.58	22.74	25.71	14.60
Kolkata	13.35	11.21	10.73	6.61	4.11
West Bengal	32.80	26.87	23.27	24.73	17.84

Source: Census of India, Various Issues.

population. The divergence between these two growth rates was really noticeable in the decade of the thirties, that is, between 1931 and 1941. Thereafter in each decade the annual growth rate of the female population has been greater than that of the male one. In the last decade the growth rates of both population have become lower than those attained in the previous one but the growth rate of female population has maintained the previous trend. In spite of the fact that the annual growth rate of female population has exceeded that of the male one, the total number of female population in each census in West Bengal has been less than that of the male one and naturally, the sex ratio in West Bengal has, instead of going in favour of the women, gone against them. We shall subsequently make a detailed discussion about the same in Chapter Ten.

Table 9.1 gives us a relief in that in comparison to earlier decades the last decade of the last century had seen a relatively low growth rate of population. It is to be noted that this growth rate includes not only the natural growth rate of population of West Bengal but also takes account of the influx of population coming from neighbouring states. Hence it will be worthwhile to concentrate on the growth of population of all districts (eighteen, considering Midnapore to be a single district) of West Bengal.

Population of West Bengal at the District Level

To make a thorough study of the growth of population of West Bengal we should portray the picture of population of this State at the district level also. At present in West Bengal there are nineteen districts whereas at the time of independence there were sixteen districts. Midnapore, the birth place of Pundit Ishwar Chandra Vidyasagar and a historic place for intensifying the freedom movement of India, is the largest district of the State. In fact, by population size, this district heads the list of all districts of India in 2001. This district occupies 15.87 percent of total land area of the State and here live 12.02 percent of population of the State. In December 2001 this district had been divided into two districts, namely (1) Midnapore East and (2) Midnapore West. Because of administrative convenience, the district of 24-Parganas had been divided into North 24-Parganas and South 24-Parganas on the first day of March, 1986. It may be mentioned that the district of North 24-Parganas, by population size, is the second top district in India in 2001. In the same way the district of Dinajpur had been subdivided into North Dinajpur and South Dinajpur in 1998.

We shall now present the annual compound growth rate of population of West Bengal and all the constituent districts per decade over the last 100 years between 1901 and 2001. This has been shown in Tables 9.4(a) to 9.4(d).

Table 9.4(c): Composition of Population of West Bengal at the District Level in 1991 (unit: number)

Districts	Male	Female	Total
Darjeeling	679323	620596	1299919
Jalpaiguri	1453194	1347349	2800543
Cooch Behar	1122306	1048839	2171145
North Dinajpur	987771	909274	1897045
South Dinajpur	632969	597639	1230608
Maldah	1360541	1276491	2637032
Murshidabad	2439342	2300807	4740149
Birbhum	1313285	1242379	2555664
Burdwan	3186833	2863772	6050605
Nadia	1989841	1862256	3852097
North 24-Parganas	3818197	3463684	7281881
South 24-Parganas	2962214	2752816	5715030
Hooghly	2271792	2083438	4355230
Bankura	1437515	1367550	2805065
Purulia	1142771	1081806	2224577
Midnapore	4284954	4046958	8331912
Howrah	1982457	1747187	3729644
Kolkata	2445328	1954491	4399819

Source: Census of India, Various Issues.

Considering the district-wise growth rate of population of West Bengal between 1991 and 2001 (vide Table 9.4(d)) we see that this rate is maximum (2.56%) in North Dinajpur and minimum

(0.40%) in Kolkata. After North Dinajpur come the districts of Maldah (2.24%), Murshidabad (2.15%), Darjeeling (2.14%), North 24-Parganas (2.06%), South Dinajpur (2.02%), Jalpaiguri (1.97%), South 24-Parganas (1.92%), Nadia (1.80%) and Birbhum (1.66%). Hence the annual growth rate of population of as many as ten of the eighteen districts of West Bengal between 1991 and 2001 has been higher than the state average of 1.65%. This leads us to conclude that for West Bengal between the last two censuses the districts lying on the border have seen a higher rate of growth of population—infiltration may be the reason behind it!

Table 9.4(d): Composition of Population of West Bengal at the District Level in 2001 and the Annual Growth Rate of Total Population (in percentage) Between 1991 and 2001

Districts	Population (number)			Percentage annual growth rate
	Male	Female	Total	
Darjeeling	826334	779566	1605900	2.14
Jalpaiguri	1753278	1649926	3403204	1.97
Cooch Behar	1271715	1206565	2478280	1.33
North Dinajpur	1260747	1181077	2441824	2.56
South Dinajpur	770443	732204	1502647	2.02
Maldah	1689409	1600751	3290160	2.24
Murshidabad	3004385	2859332	5863717	2.15
Birbhum	1545765	1466781	3012546	1.66
Burdwan	3602675	3317023	6919698	1.35
Nadia	2365054	2238702	4603756	1.80
North 24-Parganas	4635262	4295033	8930295	2.06
South 24-Parganas	3564241	3344774	6909015	1.92
Hooghly	2588322	2451725	5040047	1.47
Bankura	1634561	1557261	3191822	1.30
Purulia	1298079	1237154	2535233	1.32
Midnapore	4929000	4709473	9638473	1.47
Howrah	2242395	2031615	4274010	1.37
Kolkata	2506029	2074515	4580544	0.40

Source: Computed from census data.

Density of Population of West Bengal at the District Level

To substantiate the inference arrived at in the last paragraph let us insert another Table (Table 9.5) indicating the density of population at the district level.

An indicator of concentration of population of a country is the density of population, which is the number of people living per square kilometer area. Till 1981 Kerala had the highest density of population. But in 1991 and also in 2001 West Bengal had attained that distinction. Hence we should make a brief discussion about the density of population as revealed by the districts of West Bengal, which has been inserted in Table 9.5. From the table we notice that in both 1991 and 2001 the density of population was the highest in Kolkata, the nerve-center of West Bengal. Naturally, Howrah followed Kolkata. In 1991 there were eleven districts, namely Purulia, Bankura, Darjeeling, Jalpaiguri, South Dinajpur, Birbhum, South 24-Parganas, Midnapore, Cooch Behar, North Dinajpur and Maldah where density of population was less than the state figure of 767. In 2001 it was the same eleven districts (though there was a slight variation in rank) where density of population was less than the state figure of 904.

From the sixth column of Table 9.5 we see that the maximum change in density in 2001 was shown by North Dinajpur followed by Maldah, Murshidabad, Darjeeling, North 24-Parganas, South Dinajpur, Jalpaiguri, South 24-Paraganas, Nadia, Birbhum, Hooghly, Midnapore, Burdwan, Cooch Behar, Purulia, Bankura, Howrah and Kolkata. We may also note that there are significant gaps between ranks presented in columns (5) and (7). For example, so far as density is concerned Kolkata and Howrah had respectively the first and the second rank, but they had attained the 18th and 17th rank so far as the percentage change of density is considered. Spearmen's rank correlation coefficient between these two types of ranks in 2001 had been found to be -0.08.

The above analysis and the information presented in Table 9.5, strengthens our previous conclusion: The districts lying on the border had seen higher percentage change in density. Thus we may conclude that in the last decade of the twentieth century though

Table 9.5: Density of Population in the Districts of West Bengal in 1991 and 2001 and Ranks Attained by the Districts

Districts	Year				Percentage change of density in 2001and ranks in such changes
	1991		2001		
	Density	Rank	Density	Rank	
Darjeeling	413	16	510	16	23.49 (4)
Jalpaiguri	450	15	547	15	21.56 (7)
Cooch Behar	641	9	732	10	14.20 (14)
North Dinajpur	604	10	778	9	28.81 (1)
South Dinajpur	555	14	677	13	21.98 (6)
Maldah	706	8	881	8	24.79 (2)
Murshidabad	890	6	1101	6	23.71 (3)
Birbhum	562	13	663	14	17.97 (10)
Burdwan	861	7	985	7	14.40 (13)
Nadia	981	5	1172	5	19.47 (9)
North 24- Parganas	1779	3	2181	3	22.69 (5)
South 24- Parganas	574	12	694	12	20.91 (8)
Hooghly	1383	4	1601	4	15.76 (11)
Bankura	408	17	464	17	13.72 (16)
Purulia	355	18	405	18	14.08 (15)
Midnapore	592	11	685	11	15.71 (12)
Howrah	2542	2	2913	2	12.82 (17)
Kolkata	23783	1	24760	1	4.11 (18)
West Bengal	767	-	904	-	-

Source: Census of India, 1991, Series 26 and Census of India, 2001 Series 20, West Bengal.

the annual rate of growth of population in the State of West Bengal had, to a good extent, declined, the districts on the border had experienced a higher rate of growth of population. Unless the infiltration is stopped, West Bengal will face serious consequences of explosive population in the coming years.

Percentage Distribution of Population and Land Area at the District Level

We may now discuss the percentage distribution of population of West Bengal at the district level and the corresponding percentage distribution of land area possessed by the districts, as have been revealed through the census of 2001. These matters have been put in Table 9.6 where besides indicating these two variables, the ranks attained by the eighteen districts in respect of them have been recorded.

Table 9.6: Percentage Distribution of Population and Land Area of West Bengal Around Districts and Ranks Attained by them in 2001

Districts	Percentage of people	Rank	Percentage of Land Area	Rank
Darjeeling	2.00	17	3.55	13.5
Jalpaiguri	4.24	10	7.02	6
Cooch Bihar	3.09	15	3.82	12
North Dinajpur	3.04	16	3.58	15
South Dinajpur	1.87	18	2.50	16
Maldah	4.10	11	4.21	11
Murshidabad	7.31	5	6.00	7
Birbhum	3.76	13	5.12	8
Burdwan	8.63	3	7.91	3
Nadia	5.74	7	4.42	10
North 24-Parganas	11.13	2	4.61	9
South 24-Parganas	8.61	4	11.22	2
Hooghly	6.28	6	3.55	13.5
Bankura	3.98	12	7.75	4
Purulia	3.16	14	7.05	5
Midnapore	12.02	1	15.87	1
Howrah	5.33	9	1.65	17
Kolkata	5.71	8	0.21	18
All West Bengal	100.00	-	100.00	-

Source: Census of India, Various Issues.

Table 9.6 indicates that sharing about 16 percentage of land area of the State, the district of Midnapore has become the largest district which nourishes about 12 percentage of total population of the State. Naturally, Midnapore has topped the list in respect of both the variables studied in this section. The difference between these two types of ranks has become maximum (ten) in case of Kolkata which sharing only about 0.2 percentage of land area, has been bringing up about 6 percentage of population living in the State. The district of Howrah faces the similar situation as Kolkata where on 1.65 percentage share of the land area of the State live more than 5 percentage of the total population. Naturally, the percentage share of land area and that of population under the districts of West Bengal have not been even or egalitarian and this can not be expected also because people will like to concentrate in those places wherever they think it worth-while to live. The Spearman's Rank Correlation Coefficient between these two types of variables has been found to be +0.49, indicating that the relation between them, though positive, is not strong enough.

Distribution of Scheduled Caste and Scheduled Tribe People of West Bengal at the District Level

The people belonging to the Scheduled Caste (SC) and the Scheduled Tribe (ST) form a significant portion of people of West Bengal. According to the censuses of 1971, 1981, 1991and 2001 the share of these people in the total population of the State have respectively been 25.61, 27.61, 29.21 and 31.84 percentage. Again, between 1971 and 1981 the annual rate of growth of people belonging to the Scheduled Caste community had been 3.13 percent, though the overall growth rate of population of West Bengal had been less by one percentage (the actual figure being 2.11 percent). However, in this decade the annual growth rate of Scheduled Tribe people had been 1.94 percent. In the next decade, that is, between 1981 and 1991 the annual growth rate of people forming the Scheduled Caste community had marginally fallen to 2.97 percent whereas that of the Scheduled Tribe people had increased to 2.18 percent (an increase of 0.24 percent). It may be mentioned that in this decade the overall growth rate of people of

West Bengal was 2.23 percent.

Let us now briefly touch upon the distribution of people of these two communities in West Bengal at the district level, which has been reflected in Table 9.7.

Table 9.7: Distribution of Scheduled Caste and Scheduled Tribe people in the districts of West Bengal in 1991

Districts	Percentage of Scheduled Caste people	Rank	Percentage of Scheduled Tribe people	Rank
Darjeeling	1.31	17	4.70	7
Jalpaiguri	6.44	8	15.47	2
Cooch Bihar	6.99	5	0.35	15
Dinajpur	5.64	9	8.07	5
Maldah	2.91	14	4.50	10
Murshidabad	3.95	12	1.61	14
Birbhum	4.88	11	4.66	8
Burdwan	10.33	2	9.87	4
Nadia	6.95	6	2.38	12
North 24-Parganas	9.73	3	4.46	11
South 24-Parganas	12.24	1	1.85	13
Hooghly	6.53	7	4.63	9
Bankura	5.47	10	7.61	6
Purulia	2.64	15	11.23	3
Midnapore	8.47	4	18.12	1
Howrah	3.66	13	0.26	16
Kolkata	1.76	16	0.23	17

Source: Calculated from Census data

A study of Table 9.7 enlightens us that there is a wide variability of the percentage of Scheduled Caste and Scheduled Tribe people living in the districts. For example, in 1991, 12.24 percent of the people of the former community used to live in South 24-Parganas, 10.36 percent of those used to live in Burdwan and 9.73 percent of such people lived in North 24-Parganas. But in the same year 18.11 percent of the Scheduled Tribe people lived in

Midnapore, 15.47 percent of them lived in Jalpaiguri and 11.23 percent of them lived in Purulia. The table also demonstrates that only at Burdwan there was a high concentration of people of both the communities. The Spearman's Rank Correlation Coefficient between the ranks attained by these two communities in the seventeen districts has been very poor, that is, +0.13, indicating that there is no consistency between these two types of ranks.

Spread of Education in West Bengal

It is a fact that education plays a significant role for the furtherance of self. It has a great contribution to the family life, social life and also to economic development of any country. Swami Vivekananda had rightly said, "Education is the manifestation of perfection already in man". Investment is a central factor in the acceleration of economic growth, which leads to increasing amounts of consumption goods and services gradually being made available. It is important to note that investment consists not only of improving and adding to buildings, machinery and stocks of materials (the three principal components of physical capital) for production, but also through education and training of improving the quality of people. Here lies the importance of education. Hence we should have some knowledge about the dissemination of education among the people of the State of West Bengal.

After Independence there has been a good progress of education in this state. But, according to the latest statistics, we have been lagging very much behind as such statistics show that 69.22 percent of people of this state are literate, the corresponding figure for the all-India level being 65.38 percent. Kerala with the literacy rate of 90.92 percent (for males it is 94.2 percent while for females it is 87.86 percent) has topped the list of States and Union Territories in the sphere of education. The States and Union Territories being more advanced than West Bengal in the field of education are:

(1) Kerala (90.92), (2) Mizoram (88.49), (3) Lakshadwip (87.52), (4) Goa (82.32), (5) Delhi (81.82), (6) Chandigarh (81.76), (7) Pondicherry (81.49), (8) Andaman & Nicobar Islands (81.18), (9) Daman & Diu (81.09), (10) Maharashtra (77.27), (11) Himachal

Pradesh (77.13), (12) Tripura (73.66), (13) Tamil Nadu (73.47), (14) Uttaranchal (72.28), (15) Gujarat (69.97), (16) Punjab (69.95), (17) Arunachal Pradesh (69.68) and (17) Sikkim (69.68).

Hence the progress of as many as eighteen States and Union Territories is much superior to that of West Bengal though the literacy rate of this State is greater than the All-India average rate.

Let us now make a brief discussion about the progress of the districts of West Bengal towards spread of education. Table 9.8 contains such an information.

Table 9.8: Spread of Education Across Districts of West Bengal in 1991 and 2001 and Ranks

Districts	1991		2001	
	Literacy rate (percentage)	Rank	Literacy rate (percentage)	Rank
Darjeeling	57.95	7	72.87	6
Jalpaiguri	45.09	13	63.62	13
Cooch Bihar	45.78	12	67.21	9
North Dinajpur	34.58	15	48.63	18
South Dinajpur	46.40	15	64.46	11
Maldah	35.62	17	50.71	17
Murshidabad	38.28	16	55.05	16
Birbhum	48.56	11	62.16	14
Burdwan	61.88	6	71.00	7
Nadia	52.53	9	66.55	10
North 24-Parganas	66.81	5	78.49	2
South 24-Parganas	55.10	8	70.16	8
Hooghly	66.78	4	75.59	4
Bankura	52.04	10	63.84	12
Purulia	43.29	14	56.14	15
Midnapore	69.32	3	75.17	5
Howrah	67.62	2	77.64	3
Kolkata	70.18	1	81.31	1
All West Bengal	57.70	-	69.22	-

Source: Census of India, Various Issues. Ranks have been calculated from the literacy rate collected from this source.

Table 9.8 brings out the literacy rate attained by the districts of West Bengal in 1991 and 2001, besides indicating the ranks achieved by the districts in this sphere in both the years. In 1991, the overall literacy rate in the State was 57.70 percent while in 2001 this rate improved to 69.22. It is clear from this table that in comparison to 1991 the literacy rate in 2001 has been much higher in every district. In 1991, Kolkata was the only district to have attained the literacy rate much higher than the state level whereas in 2001 Kolkata, North 24-Parganas, Howrah, Hooghly, Midnapore, Darjeeling, Burdwan and South 24-Parganas had attained the literacy rate higher than the state average. This implies that with the passage of time education in West Bengal instead of being centralized, has become decentralized. This is, undoubtedly, a great achievement. It can be noticed from the table that the difference of the ranks attained by the districts in 1991 and 2001 are not so much pronounced and consequently, Spearman's rank correlation coefficient has become +0.94, a high figure. The table also shows that Maldah attained the poorest performance in literacy rate in both the years. North 24-Parganas achieved a spectacular performance in the field of education as here the literacy rate had moved from 56.7 percent in 1991 to 78.49 percent in 2001. It may be mentioned here that in the sphere of education the attainment of the capital of West Bengal is tantamount to that of the capital of India.

Conclusion

After analysing the various aspects of the growth of population in the State of West Bengal we can arrive at the following conclusions:

(1) The annual rate of growth of population of West Bengal between 1991 and 2001 has declined to a good extent. The population of West Bengal at the start of the next census in 2011, will be about 95 million, assuming that the present trend rate of growth of population will prevail.

(2) The districts of West Bengal nearing border have faced higher rate of growth of population.

(3) The districts of West Bengal in the neighbourhood of borders have also experienced greater rate of change in the density of population.

(4) The share of Scheduled Caste and Scheduled Tribe people in total population of West Bengal has, with the passage of time, been increasing.

(5) After Independence West Bengal has attained good success towards the spread of education, but in comparison to other states we have much to attain.

REFERENCES

1. Basu, Kalipada (1989): *West Bengal Economy, Past, Present and Future.*
2. Bhattacharya, Dhires (ed) (1972): *Focus on West Bengal: Problems and Prospects*, Samatat Prakashani, Kolkata
3. Census of India, Various Issues.
4. *Economic Review*, Govt. of West Bengal, Various Issues.
5. *Economic Survey*, Govt. of India, Various Issues.
6. Konar, D.N. (2000): "Development experience in Eastern Idnia," *20th Annual Conference Volume of the Bangiya Arthaniti Parishd.*
7. Konar, D.N. (2004): "Pattern of Development in Eastern india" in the book (ed) *Eastern India, The Emeging Issues in Development*, Published by the Department of Economics, Rishi Bankim Chandra College, Naihati, 24-Parganas (N), West Bengal.

10

SEX RATIO IN WEST BENGAL

We may now explain the pattern of sex ratio in West Bengal, which has been reflected in Table 10.1 and it makes it clear that in this state the sex ratio was always unfavourable to women. There were 945 women against 1000 men in 1901 but in 1941 there became 852 women per 1000 men. So this ratio became peak in 1901 while it became the least in 1941. From 1951 it started improving and it became 934 in the current census when it had just crossed the all-India figure of 933. Looking at the urban trend we observe that it was always highly adverse to women, more so in the pre-independent period when it had fallen from 652 in 1901 to 559 in 1941. In 1951 there was a big jump and from then it had started increasing from 660 in 1951 to 893 in 2001. The rural scenario, in West Bengal, was also unkindful to the women. In 1901 it was really favourable to women, but in the subsequent censuses it turned unfavourable and this unfavourableness persisted till 1991. It is only the current census when there was an advancement of ten points, from 940 to 950 though it failed to touch the initial figure of the earlier century.

For a quick understanding of the trend of sex ratio in West Bengal (Total, Urban and Rural) and India we have used a diagram (vide Diagram 10.1) which is self-explanatory. We see from the diagram that the sex ratio in Urban Bengal has been increasing while that in rural Bengal has remained stagnant at 950. The sex ratio in India and West Bengal in 1941 had became identical and then onward the ratio of India was almost on the track of that of West Bengal.

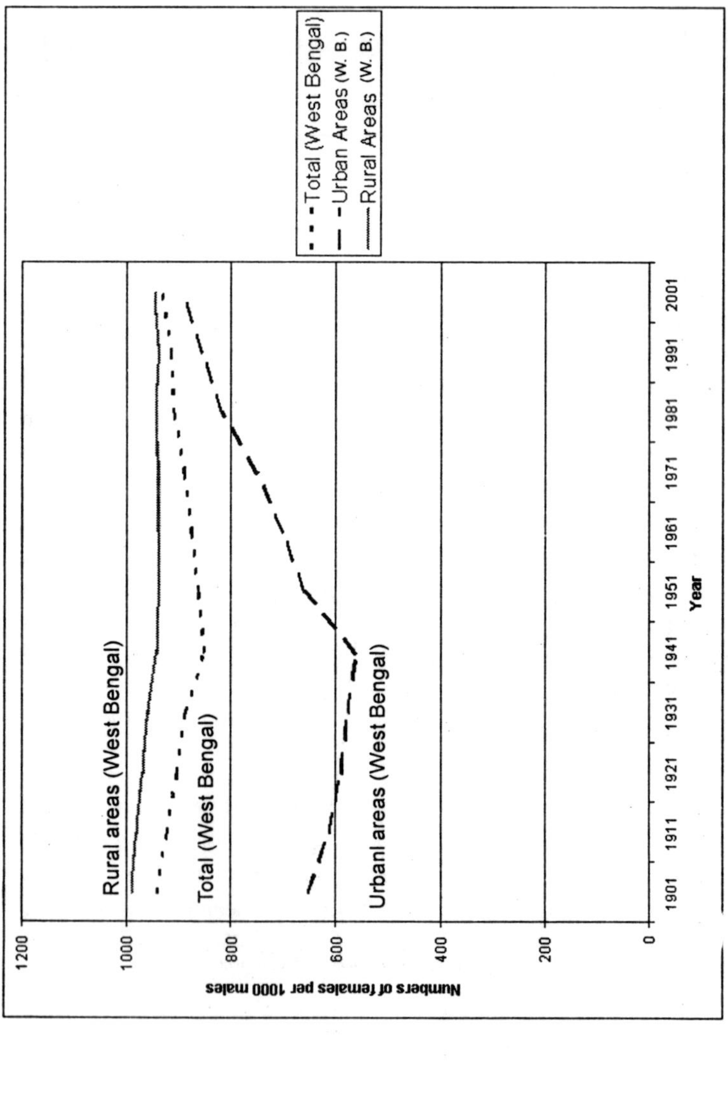

Diagram 10.1: Diagrammatic Presentation of Sex-Ratio in West Bengal and India

Let us now present the district-wise sex ratio in West Bengal in 1991 and 2001 as shown in Table 10.2 which demonstrates that in the district-level the ratio does not favour the ladies who have become the worst sufferer in the capital of West Bengal itself. Next to Kolkata comes Howrah. To display the nature of female-male ratios across districts of West Bengal in 1991 and 2001 we have utilized Table 10.2 which will amply reveal that in comparison to 1991 the ratio in 2001 had, in general, become congenial to the females. For example, in 1991 Bankura was the solitary district to have the ratio more than 950 while in 2001 besides Bankura, Midnapore, Purulia, Murshidabad and South Dinajpur showed such a trend. In 1991 there were three districts showing adverse ratio while in 2001 the number of such districts got reduced to only one. In the last two censuses we have given ranks to each of eighteen districts of West Bengal in the usual manner and computed Spearman's rank correlation coefficient which has been found to be +0.92, a very high value. This suggests that the pattern of sex ratio as revealed by the districts of West Bengal in the last two censuses has remained unchanged.

Table 10.1: Female-Male Ratio in West Bengal in the Census Years between 1901 and 2001

Year	Total	Urban Areas	Rural Areas
1901	945	652	994
1911	925	614	982
1921	905	591	971
1931	890	578	961
1941	852	559	945
1951	865	660	939
1961	878	701	943
1971	891	751	942
1981	911	819	947
1991	917	858	940
2001	934	893	950

Source: Census of India, Various Issues.

We have already seen that a very peculiar aspect of the census 2001 is the highly declining gender ratio in the age group 0-6 years. We now intend to demonstrate how far the districts of West Bengal are responsive to this aspect. Such responsiveness will appear in Table 10.3 where we have presented the girl-boy ratios in the age group 0-6 across the districts of West Bengal in 1991 and 2001. We have also given normal ranks to the districts.

Table 10.2: Female-Male Ratio and Ranks Across Districts of West Bengal in 1991 and 2001

Districts	1991		2001	
	Ratio	Rank	Ratio	Rank
Darjeeling	914	14	943	11
Jalpaiguri	927	11	941	12
Cooch Bihar	935	9	949	6.5
North Dinajpur	921	12	937	14
South Dinajpur	944	4.5	950	5
Maldah	938	7	948	8
Murshidabad	943	6	952	4
Nadia	936	8	947	9.5
North 24-Parganas	907	15	927	15
South 24-Parganas	929	10	938	13
Kolkata	799	18	828	18
Howrah	881	17	906	17
Hooghly	917	13	947	9.5
Burdwan	899	16	921	16
Birbhum	946	3	949	6.5
Bankura	951	1	953	2.5
Midnapore	944	4.5	955	1
Purulia	947	2	953	2.5

Spearman's Rank Correlation Coefficient has been +0.92.
Source: Census of India 2001, Series 20; West Bengal.

From Table 10.3 we note that at all-Bengal level this ratio in 2001 had reduced by eighteen points. Not only that, the mean ratio in West Bengal in 2001 had declined to 963.2 from 967.9 attained a decade earlier. Besides, the standard deviation of the said ratio in 2001 had increased to 12.1 from 8.3 attained in 1991. These results are really a matter of grave concern. We also note that for this ratio, Spearman's rank correlation has been +0.90, almost equal to the earlier value. So in respect of gender ratio in the age group 0-6 the pattern as revealed by the districts of West Bengal in 1991 and 2001 has almost remained unchanged.

Table 10.3: Girl-Boy Ratios and Ranks in the Age Group 0-6 Across Districts of West Bengal in 1991 and 2001

Districts	1991		2001		Change, Increase(+) or Decrease (-)
	Ratio	Rank	Ratio	Rank	
Darjeeling	976	3.5	971	5	-5
Jalpaiguri	973	7.5	973	3.5	0
Cooch Bihar	967	11	968	8.5	+1
North Dinajpur	974	5.5	973	3.5	-1
South Dinajpur	974	5.5	968	8.5	-6
Maldah	960	14.5	967	10.5	+7
Murshidabad	977	2	975	1.5	-2
Nadia	983	1	975	1.5	-8
North 24-Parganas	969	9.5	961	12	-8
South 24-Parganas	973	7.5	969	6.5	-4
Kolkata	955	17	923	18	-32
Howrah	962	12.5	959	14	-3
Hooghly	960	14.5	952	16.5	-8
Burdwan	959	16	960	13	+1
Birbhum	976	3.5	969	6.5	-7
Bankura	962	12.5	955	15	-7
Midnapore	953	18	952	16.5	-1
Purulia	969	9.5	967	10.5	-2
West Bengal	967	-	963	-	-4
India	945	-	927	-	-18

Source: Census of India, 2001, Series 20, West Bengal.
Spearman's Rank Correlation Coefficient is +0.90.

A SEARCH FOR UNFAVOURABLE SEX RATIO

A minute observation on the diagram presenting the sex ratio in India and West Bengal reveals that the general trend of sex ratio in West Bengal is a declining one between 1901 and 1941 and then an increasing trend starting from 1951 has gone uninterruptedly. Similar is the pattern of the urban ratio which had continuously declined between 1901 and 1941 and thereafter it went on increasingly. However, in case of rural areas there is more or less a declining trend. Let us now try to explain the reasons behind such a pattern.

Reasons Behind Low Sex Ratio Between 1901 And 1941

A sex ratio unfavourable to women does not support the normal situation which, in all possibility, divulges that death rate among females compared to those of males is low because of biological reasons. It enables females to resist diseases more effectively than males, as is also the fact that males more than females are confronted with strains and stresses of life. But still then we face in Bengal and in India as well a sex ratio which has a preponderance of males.

The explanation behind the strange nature of sex ratio in West Bengal requires a thorough study for a sharply downhill trend between 1901 and 1941 and then an upward trend from 1951. Several reasons may be provided to account for such a behaviour of sex ratio specially in the initial stage. There may not, however, be unanimity in these explanations. Still then let us make a modest attempt to search for some causes leading to such behaviour of sex ratio till 1941.

One important reason behind the sex ratio being unfavourable to women in West Bengal is that the average number of female births to one thousand male births is relatively far low. For example, in 1961 census the general sex ratio in West Bengal was 878 but in that census 923 female births were recorded.

A second possible reason for the downward movement of the said ratio till 1941 may be found in the "under-reporting of the female children and the inclusion of fictitious male children"

(Hicks, 1984). The prevalence of polygamy in both the Hindu and Muslim families led many husbands to understate the number of their wives. The under-reporting of their females was, in all possibility, a regular phenomenon because of the extremely poor status of the women at that time. Professor A. K. Sen (1983) rightly observed, "the general position of women in Indian society is nothing short of scandalous". The social and economic atmosphere of the then society compelled the heads of so many families to understate the number of females in their families. It is also pointed out that in the Indian censuses there has been under-enumeration. "There is suspected under-enumeration in the population in the census of 1931 mainly in Gujarat due to non-co-operation movement and over-enumeration in 1941 census in the Punjab and Bengal" (Mukerjee, 1982). Some may, however, point out that under-enumeration or over-enumeration whatever may be the case, may be common to both males and females.

A third possible reason for the declining trend in the sex ratio lies in the easy mobility of the males in and around the place of employment. The high general demand for labor in the industrial belts of West Bengal, specially in Kolkata, 24-Parganas, Howrah and Hooghly and the increasing demand for labor in the tea gardens in the Dooars regions led many males from adjoining states to shift to these regions. In the process they had left their womenfolk in the domicile for looking after agriculture, animal husbandry and the like. The non-availability of adequate gainful employment for females discouraged them to shift to urban areas, while the male folk found themselves confident to face the hazards of urban life. This factor alone is sufficient to explain the severely unfavourable sex ratio in Kolkata till 1941.

The fourth and the most fundamental reason behind a very low sex ratio may be found in the poverty of the Indian people. It is this acute poverty that makes a certain amount of bias in the allocation of many essential items, viz., medicines, health attention, food etc. The preference for providing good food and medicine to the breadwinner is the manifestation of the much-sought-after modicum of economic security that the breadwinner provides. It is because of this reason that girls, in most cases, in India are tolerated rather than welcome–they are looked after as unwanted children. Male children, on the other hand, are brought up with utmost care as they are supposed to help the parents. The economic considerations together with the religious belief that "male children will perform the funeral rites for the parents" are strong enough to make them biased towards the male children.

One likely consequence of extreme poverty is the early marriages of the girls, the possible effects of which may manifest themselves in the relatively large number of death among women in the reproductive ages, fifteen to fortynine, due partly to inexperience to grapple with the problems of malnutrition and partly due to physical incapacity to bear the burden of motherhood. This, along with extremely low or non-existent medical facilities and the absence of pre-natal and post-natal care, account for the high maternal mortality rate in many parts of India. Also is the fact that infant mortality among girls is very high.

Reasons Behind The Upward Trend Of The Sex Ratio From 1951

A fundamental cause behind the upward trend of the sex ratio from 1951 may be found in the increase in the life expectancy of the people in India specially, from 1951. In the period between 1901 and 1910 it was only 23 years. In the next decade it was estimated to be even lower at 20.2 years. It started rising, though slowly, after 1921. The rise became substantial from 1951. It was 41.3 years in 1961, 45.6 years in 1971 and 54.4 years in 1981. The life expectancy of females in India increased from 31.7 years in 1951 to 44.7 years in 1971 and further to 54.7 years in 1981. Thus in the decade from 1971, the female life expectancy had improved by 10 years and had overtaken the male life expectancy which was 54.1 years in 1981.

Another vital cause in favour of the increase in sex ratio in West Bengal after 1941 may be the in-migration of males as well as females to tea plantations and as labourers to the industrial areas of Kolkata, 24-Parganas, Howrah and Hooghly. Before independence the labor class from States like Bihar, Orissa and Uttar Pradesh used to come to Bengal in search of employment, leaving their families in the domicile. But after partition this class of people found it convenient and economic as well to transfer their families in the vicinity of the place of employment. Besides, the trade and commerce class from various parts of India used to arrive at Bengal, specially in the industrial and business centers, keeping their near and dear ones at home. So long they had hesitation at deciding whether they would permanently settle down. But with the dawn of independence and with the onset of economic and political stability after partition, this class also found it worthwhile to reside in the tranquil Bengal permanently. Nourishing this idea, this class of people started purchasing assets and they shifted their families from their permanent place of residence to different parts of Bengal. Together with this in-migration, the people (the educated and the financially better off) from the interior parts of the rural areas have started coming and residing in the fringes of the urban areas. This trend is itself a reason for making the rural sex ratio unfavourable to women whereas the urban sex ratio favourable to them. Again, for the last few years of the twentieth century construction is booming in the urban areas of West Bengal. In this construction industry a large number of hard-working females coming principally from Bihar, Andhra Pradesh and the interior parts of Bengal are employed. They also contribute to favour the urban sex ratio.

Thirdly, with the spread of education and as a result of rising consciousness of the females, the marriage age has, to a great degree, increased. This has, obviously, lowered the maternal mortality rate. Again, mobility among females has significantly increased now-a-days: neither they are only brides of the families serving fathers-in-law, mothers-in-law, husbands and the like with cooked food and drink; nor are they the mere spectators of ineffable torture inflicted upon them. They are as good as their opposites. They also struggle for existence and such feeling has made them to move to and fro in search of a job, security and betterment of self. This factor is, to a good degree, held responsible for making the rural sex ratio unfavourable.

Another probable reason in support of the upward shift in the sex ratio from 1951 is the influx of displaced population after the partition. From different sources it is known that after partition at least six million people moved to West Bengal from East Bengal, now Bangladesh. It is generally reported that among the people coming from East Bengal there are more females than males and most of the displaced persons take shelter in the urban areas where there are so many attractions and advantages. Such an influx of population has added more to women than to men. This factor is, to a large extent, held responsible for improving the sex ratio in West Bengal from 1951 in general, and making a drastic change in the urban sex ratio, in particular.

Moreover, the government both at the center and at the state had made sustained efforts to promote family planning programs. Together with the introduction of these programs more attention has been paid towards family welfare programmes which have reduced the death rates of the females. This means that direct involvement of the state has led to an upward rise in the sex ratio.

Finally, the dissemination of education and modern outlook in the far and wide of the country, the enjoyment of equal right in casting a vote, the frequent appearance in the polling centers for exercising a vote in various elections, the adoption of so many precautionary measures for making the census much more intensive and scientific, the influence of the Hindu and Muslim Marriage Acts, the growing status of the women in the post- independent India and more recently, the parents' satisfaction with having a child irrespective of the sex–all these have led to the proper reporting of the sex and have also made the demographic gap between males and females narrower and narrower.

CONCLUSION

From the census data it is vividly clear that there has been an improvement in the number of females per one thousand of males and this heralds an increase in the sex ratio in favour of the females. The increase in the sex ratio is very significant in that it is indicative of the high status of women in the society. It also signifies dynamism among the female population in the community. The sex ratio is, to an extent, an index of the stability of the urban population. So an increasing urban sex ratio as it has been happening in West Bengal, indicates that the urban people of Bengal are becoming much more stabilized so far as their family life is considered. But together with this optimism we should not forget a pessimistic picture that has cropped up with the amniocentesis tests which were actually conceived as a means of discovering whether the child in the womb suffered from any genetic defects so grave as to warrant an abortion. But these tests are becoming much popular in India only as an instrument for identifying the sex of the child in the womb. In case the child is detected to be a female, there is a general tendency among the mothers to go in for an abortion. It is reported from Chatterjee (1987) that between 1979 and 1982, 78,000 female foetuses were aborted by couples who had volunteered for the tests. So amniocentesis tests followed by abortion of female foetuses will, far from strengthening the sex ratio, weaken it and the demographic gap between males and females will widen.

It is heartening to state that from January 1996 the Government of India had legally forbidden the determination of sex of an unborn foetus by pre-natal diagnostic technique. But in spite of the legal ban abortion of the female foetus is going on. We have already stated that a survey of a clinic at Mumbai revealed 7999 of the 8000 aborted foetuses were females. To get rid of such a sorry state of affair what we need is mass awareness, which alone can prevent the said abortions. We may rightly say, "Legislation will mean nothing without much awareness".

REFERENCES

1. Census of India: Various Issues
2. Census of India, 2001, Series 20, West Bengal.
3. Chatterjee, Shoma A (1987): "Foetal Guillotine". *The Sunday Magazine*, The Amritabazar Patrika, March 8.
4. Hicks, Mukherjee & Ghosh (1984): *The Framework of the Indian Economy*, Oxford University Press.

5. Konar, D. N. (1990): "An Attempt to Analyse the Trend of Sex Ratio in West Bengal." The Economic Studies, Kolkata, Vol. 28, No. 3.

6. Konar, D. N. (1998): "Unfavourable Sex Ratio and the Deprivation of Women in India". *Artha Beekshan*, Vol. 7, No. 2, December.

7. Mukherjee, S (1982): "Under-Enumeration in Indian Census—Impact on Intra-Censal Population Growth: 1901-81", *Economic and Political Weekly*, November 13,20.

8. Sen, A. K. (1983): "The Doing and the Undoing", *Economic and Political Weekly*, February 12.

9. Sen, A. K. (1988): A Lecture on "Economics, Culture and Sex Differential in Mortality" at the *71st Annual Conference of the Indian Economic Association* at Jadavpur University, Kolkata, on 31st December.

11

IN SEARCH OF THE DEPRESSED AREAS OF WEST BENGAL

INTRODUCTION

The State of West Bengal coming into existence with the attainment of independence by India in August 1947 epitomizes through its districts and regions an unequal pattern of development, which is also visible all over India. The economy of West Bengal is basically rural-based. This is established by the fact that in 1991 and 2001 respectively 72.52 percent and 71.97 percent of her people lived in rural areas and still now (1999-2000) nearly 31 percent of the state's Net State Domestic Product is reaped through agriculture and allied services. However, historical and political factors have influenced the levels of industrialization and urbanization in this State. As a consequence, some areas have grown at a rapid pace and have become more developed while some others have progressed at a snail's pace and have remained virtually backward. Even in the rural scenario which is more traditional and less modernized, there are vast divergences from one district to another.

With reference to the above background the topic of consideration seeks to fulfil a two-fold objective:

1) To identify which of the eighteen districts of West Bengal (the district of Midnapore has been considered to be the

single district) can be deemed to be depressed in the light of some indicators noted below and

2) to see if there is any change in the ranking of the districts of West Bengal on the basis of those criteria.

DEPRESSED AREAS: THE CONCEPT AND SOME INDICATORS

It is not possible to give a unique definition of the term "depressed" areas. Even the so-called most advanced areas may be depressed from some point of view and also from some other points of view they may be deemed non-depressed. By depressed areas let us mean under-developed or backward areas. We may say that there are many areas which are handicapped by the existence of such factors as hinder these areas to have high or moderate rate of growth of per capita income, an indicator with the help of which the advancement of an area is usually measured. But the per capita income alone cannot be a meaningful symbol of detecting whether an area is progressing or not because that will represent only the entire aspect of economic growth and thus will fail to divulge fully the standard of existence of the common people. So instead of concentrating on per capita income as the solitary indicator of depression of a region let us point out some factors that lead to lower per capita income in a region. Our indicators are basically related to different aspects of population of the regions concerned because in determining the rural urban differential or inter-group differential in the levels of progress within a country population plays a very vital role. Some such indicators resulting from human aspect on the basis of which we may say whether an area may be called depressed or not are:

i) Rate of growth of population,

ii) Density of population,

iii) Rate of urbanization,

iv) Percentage share of Scheduled Caste and Scheduled Tribe people in total population,

v) Sex composition of the population,

vi) Literacy rate and

vii) Percentage of main workers in total population.

A SHORT PROFILE OF THE DISTRICTS OF WEST BENGAL ON THE BASIS OF THE ABOVE INDICATORS

A brief description of all the eighteen districts of West Bengal on the basis of the previous indicators can now be made and they can also be ranked according to those criteria. The ranking of the districts in 1991 and 2001 have been made on the basis of the available information collected from various census records and different issues of Economic Review, Government of West Bengal.

a) Rate of Growth of Population

The magnitude of population is an important factor in determining the nature of an economy. Higher growth rate of population combined with excessive concentration in rural areas points to the backward structure of an economy. We have already seen that in comparison to the decade 1971-81 and earlier decades also the growth rate of population progressively slowed down in all the districts between 1981-91 and also between 1991-2001. Hence from this aspect each and every district of this state has proceeded towards advancement.

We have classified eighteen districts of the State of West Bengal in Table 11.1 according to the decade-wise growth rate of total population in 1991 and 2001 censuses. In doing so, the district having the least growth rate has been given rank 1 while that having maximum growth rate has received rank 18 and so on. We see from the bottom of Table 11.1 that the rank correlation coefficient among the eighteen districts in census years 1991 and 2001 with regard to the growth rate of population is very high (+0.79), revealing that there is very little change in the ranks of the districts with regard to decade-wise growth rate of population even when a span of ten years has elapsed.

Table 11.1: Decade-wise Growth Rates of Population in the Districts of West Bengal during 1981-91 and 1991-2001 and Ranks Attained by the Districts from Lowest to Highest

Districts	1981-1991		1991-2001	
	Growth rate of population	Rank attained	Growth rate of population	Rank attained
Darjeeling	26.91	12	23.54	15
Jalpaiguri	24.44	9	21.52	12
Cooch Bihar	22.55	6	14.15	4
North Dinajpur	34.00	18	28.72	18
South Dinajpur	24.39	8	22.00	13
Maldah	29.78	14	24.77	17
Murshidabad	28.20	13	23.70	16
Birbhum	21.94	4	17.88	9
Burdwan	25.13	10	14.26	5
Nadia	29.95	15	19.25	10
North 24-Parganas	31.69	17	22.64	14
South 24-Parganas	30.24	16	20.89	11
Hooghly	22.43	5	15.72	8
Bankura	18.12	2	13.79	2
Purulia	20.00	3	13.96	3
Midnapore	23.57	7	15.68	7
Howrah	25.71	11	14.60	6
Kolkata	6.61	1	4.11	1
West Bengal	24.73	-	17.84	-

Rank correlation coefficient: +0.79

b) Density of Population

The density of population, that is, population per square kilometer of total area of a country may not be directly related to the furtherance of that country. But so far as West Bengal is concerned we can say with certainty that this is a very good sign of determining whether an area under discussion is depressed or not. The higher the density of population the more developed the area is going to be and the lower the density, the less developed the area. The density of population is not uniform among different parts of West Bengal. It is supreme (23783 in 1991 and 24760 in 2001) in Kolkata followed respectively by Howrah (2542 in 1991 and 2913 in 2001) and Hooghly (1383 in 1991 and 1601 in 2001). Unhesitatingly, these are the two top ranking districts in this State so far as economic development is considered. From this point of view each of the districts of West Bengal having density of population lower than the average density of population of the State (767 in 1991 and 904 in 2001) may be viewed depressed in the respective years.

Table 11.2: Density of Population in the Districts of West Bengal in 1991 and 2001 and Ranks Attained by them from Highest to Lowest

Districts	1991		2001	
	Density	Rank	Density	Rank
Darjeeling	413	16	510	16
Jalpaiguri	450	15	547	15
Cooch Bihar	641	9	732	10
North Dinajpur	604	10	778	9
South Dinajpur	555	14	677	13
Maldah	706	8	881	8
Murshidabad	890	6	1101	6
Birbhum	562	13	663	14
Burdwan	861	7	985	7
Nadia	981	5	1172	5
North 24-Parganas	1779	3	2181	3
South 24-Parganas	574	12	694	12
Hooghly	1383	4	1601	4
Bankura	408	17	464	17
Purulia	355	18	405	18
Midnapore	592	11	685	11
Howrah	2542	2	2913	2
Kolkata	23783	1	24760	1
West Bengal	767	-	904	-

Rank correlation coefficient: +0.996

The districts of West Bengal have been arranged according to their ranks in density of population in 1991 and 2001. The highest rank (1) has been provided to the district with the highest density while the lowest rank (18) has been accorded to the district with the least density. The value of the rank correlation coefficient, +0.99, reveals that the districts are almost perfectly positively correlated and this shows that even with a lapse of ten years the districts have almost failed to change their rank in this sphere of demographic variable. From this standpoint, therefore, the districts that remain depressed in 1991 still did so in 2001.

c) The Rate of Urbanization

The rate of urbanization of a country, that is, the percentage of urban population to total population is also an important symbol of judging whether an economy under review is depressed or not. In fact, the rate of urbanization itself depends upon economic development. In 1991, 27.48% of total population in this State lived in urban areas while in 2001, 28.03% of total population lived in urban areas. Kolkata is fully an urban area. In North 24-Parganas, Howrah, Burdwan, Hooghly and Darjeeling the respective percentage of total population living in urban areas in 2001 was 54.30, 50.39, 37.18, 33.48 and 32.44. This criterion thus suggests that each of these districts falls in the domain of non-depressed areas while the remaining districts where the percentage of urban population is less than the average figure of West Bengal, come in the jurisdiction of depressed areas.

Each and every district of West Bengal has been ranked in 1991 and 2001 according to the rate of urbanization. The highest rank has, obviously, gone to Kolkata, while the lowest rank (18) has been given to Maldah in both the census years. The rank correlation coefficient with regard to this variable among the eighteen districts has been found to be +0.98, suggesting that the ranking of the districts with respect to this aspect has remained almost unaltered even with the lapse of ten years. So from this criterion the districts remaining depressed in 1991 remained almost so even in 2001.

Table 11.3: Percentage of Urban Population of Total Population in each district of West Bengal and Ranks Attained by them from Maximum to Minimum

Districts	1991		2001	
	Rate of urbanization	Rank	Rate of urbanization	Rank
Darjeeling	30.47	6	32.44	6
Jalpaiguri	16.36	8	17.74	8
Cooch Bihar	7.81	17	9.10	15
North Dinajpur	13.34	10	12.06	12
South Dinajpur	13.35	9	13.09	10
Maldah	7.07	18	7.32	18
Murshidabad	10.43	12	12.79	11
Birbhum	8.98	15	8.58	16
Burdwan	35.09	4	37.18	4
Nadia	22.63	7	21.27	7
North 24-Parganas	51.23	2	54.30	2
South 24-Parganas	13.03	11	15.77	9
Hooghly	31.19	5	33.48	5
Bankura	8.29	16	7.37	17
Purulia	9.44	14	10.07	14
Midnapore	9.85	13	10.49	13
Howrah	49.58	3	50.39	3
Kolkata	100.00	1	100.00	1
West Bengal	27.48	-	28.03	-

Rank correlation coefficient: +0.98

d) Percentage Share of Scheduled Caste and Scheduled Tribe People in Total Population

The proportion of the persons belonging to Scheduled Caste and Scheduled Tribe is indicative of depression of a region. The larger the proportion of these people in a region the more depressed will be that region and the less this proportion in a region the more developed will be that region. From this consideration the districts of Jalpaiguri, Cooch Bihar, Bankura, North Dinajpur, South

Dinajpur, Purulia, Birbhum, South 24-Parganas, Burdwan, Darjeeling and Nadia where the total share of Scheduled Caste and Scheduled Tribe people in total population had exceeded more than the state average of 29.21 % in 1991 in the decreasing order, exhibit the symptom of backwardness. Again, by virtue of this criterion, the districts of Jalpaiguri, Cooch Bihar, Bankura, Purulia, Birbhum, South 24-Parganas, South Dinajpur, North Dinajpur, Burdwan, Nadia and Darjeeling have been deemed depressed in 2001.

Table 11.4: Share of Scheduled Caste and Scheduled Tribe People in Total Population of each District in West Bengal and the Ranks Attained from Lowest to Highest

Districts	1991		2001	
	Percentage share	Rank	Percentage share	Rank
Darjeeling	29.91	8	34.25	8
Jalpaiguri	58.03	18	59.47	18
Cooch Bihar	52.37	17	54.85	17
North Dinajpur	34.39	11	35.90	10
South Dinajpur	45.72	16	47.37	16
Maldah	24.62	5.5	25.54	4
Murshidabad	14.69	2	15.33	2
Birbhum	37.62	13	37.90	13
Burdwan	33.66	10	37.02	12
Nadia	31.36	9	34.49	9
North 24-Parganas	22.18	4	27.51	6
South 24-Parganas	35.68	12	36.14	11
Hooghly	28.17	7	31.27	7
Bankura	41.70	15	43.09	15
Purulia	38.58	14	40.00	14
Midnapore	24.62	5.5	26.65	5
Howrah	16.06	3	19.68	3
Kolkata	6.64	1	14.54	1
West Bengal	29.21	–	31.84	–

Rank correlation coefficient: +0.99

All the districts of the State of West Bengal in the census years 1991 and 2001 have been ranked according to the share of these people in total population. The highest rank, 1, has been accorded to the district having least share while the least rank, 18, has been given to the district having maximum share.

It is interesting to see that the ranking of the districts in West Bengal in 1991 and 2001 on the basis of the percentage total share of Scheduled Caste and Scheduled Tribe people in total population has remained absolutely unchanged as is revealed by the fact that the rank correlation coefficient of the districts in the last two censuses is perfectly positive (+0.99).

e) Sex Composition of the Population

The sex ratio is not a good index of either depression or prosperity of a country. But so far as our State of West Bengal is concerned we may point out with definiteness that the economic prosperity of a district has been found to be inversely related to the sex ratio: the more advanced the district, the more unfavourable the sex ratio and the more backward the district, the more favourable the sex ratio. For example, the first two most advanced districts of our State, namely Kolkata and Howrah in decreasing order, had the last two ranks viz., 18 and 17, respectively, in this attribute in both the years 1991 and 2001. There may, however, be various reasons behind this.

The relation between sex ratio and the rate of urbanization in the districts of West Bengal has been seen to be inverse. This is established by the fact that the rank correlation coefficient between the rate of urbanization and the sex ratio in all the eighteen districts in 1991 has been found to be –0.82 and that in 2001 has been –0.76. Hence, the more urbanized a district, the more unfavourable the sex ratio and the more advanced the district; on the other hand, the less urbanized the district, the less unfavourable the sex ratio and the less anvanced the district.

All the eighteen districts of the State have been ranked on the basis of sex ratio in both the years 1991 and 2001 by the usual manner. This means that the district showing the highest figure of sex ratio has been attributed rank 1 while that showing the least value of this ratio has been given rank 18 and so on. The rank correlation coefficient with regard to this ratio among the eighteen districts in 1991 and 2001 has been calculated to be +0.92. This suggests that even with the consideration of sex ratio the ranking of the districts of this State has remained, more or less, unchanged in 1991 and ten years hence.

Table 11.5: Sex Ratio in the Districts of West Bengal in 1991 and 2001 and Ranks Attained by them from Maximum to Minimum

Districts	1991		2001	
	Sex Ratio	Rank	Sex Ratio	Rank
Darjeeling	914	14	943	11
Jalpaiguri	927	11	941	12
Cooch Bihar	935	9	949	6.5
North Dinajpur	921	12	937	14
South Dinajpur	944	4.5	950	5
Maldah	938	7	948	8
Murshidabad	943	6	952	4
Birbhum	946	3	949	6.5
Burdwan	899	16	921	16
Nadia	936	8	947	9.5
North 24-Parganas	907	15	927	15
South 24-Parganas	929	10	938	13
Hooghly	917	13	947	9.5
Bankura	951	1	953	2.5
Purulia	947	2	953	2.5
Midnapore	944	4.5	955	1
Howrah	881	17	906	17
Kolkata	799	18	828	18
West Bengal	917	-	934	-

Rank correlation coefficient: +0.92

In 1991 the sex ratio of the State was 917. Hence all the districts having a figure of sex ratio less than this figure have been considered depressed: Kolkata, Howrah, Burdwan, North 24-

Parganas and Darjeeling were five such districts. On the other hand, the sex ratio of the State in 2001 was 934. Kolkata, Howrah, Burdwan, and North 24-Parganas being the four districts where sex ratio was less than the state average, were depressed in 2001.

f) Literacy Rate

The level of literacy of a country determines the quality of population of that country and thereby it affects the advancement of that country. For obvious reasons, the higher the level of literacy in a region the more forward will be that region and the lower the level of literacy the more backward will be that region. Therefore, the literacy rate, that is, the percentage of literate people out of total population of a region can, unambiguously, be an important symbol of measuring the backwardness of a region.

According to the census of 1991, 57.7% of the people of West Bengal are literate. In 1981, this figure was 47.9%. So the dissemination of education within a gap of ten years in this State has been quite praiseworthy. In 1991 the percentage of literate people in India as a whole was 51.6. So in comparison to other states West Bengal was much advanced in this sphere of education. Again, in 2001, 69.2% of the people of West Bengal was literate while at the all-India level this percentage was 65.4. Hence even in 2001, West Bengal was ahead of many states in this field.

However, the spread of education in the far and near of West Bengal has not been of a uniform nature. There are some districts where the literacy rate has been far greater than the average rate of West Bengal as a whole and there are some other districts where this rate is extremely meagre. From this point of view all the districts of West Bengal having literacy rate less than the average figure of West Bengal can be considered to be depressed. Thus in 1991 the following districts in descending order of the rate of literacy viz., South 24-Parganas (55.10), Nadia (52.53), Bankura (52.04), Birbhum (48.56), South Dinajpur (46.40), Cooch Bihar (45.78), Jalpaiguri (45.09), Purulia (43.29), Murshidabad (38.28), Maldah (35.62) and North Dinajpur (34.58) fall in the region of depression. In 2001 these districts were Cooch Bihar, Nadia, South Dinajpur, Bankura, Jalpaiguri, Birbhum, Purulia, Murshidabad, Maldah and North Dinajpur.

All the districts of West Bengal have been ranked in 1991 and 2001 on the basis of the literacy rate and the rank correlation coefficient has been found to be +0.95. This authenticates that in the sphere of spread of education there has been little change in the ranking of the districts in 2001 from what their position was in 1991.

Table 11.6: Literacy Rate in the Districts of West Bengal and Ranks Attained by the Districts from Highest to Lowest

Districts	1991		2001	
	Literacy Rate (%)	Rank attained	Literacy Rate (%)	Rank attained
Darjeeling	57.95	7	72.87	6
Jalpaiguri	45.09	14	63.62	13
Cooch Bihar	45.78	13	67.21	9
North Dinajpur	34.58	18	48.63	18
South Dinajpur	46.40	12	64.46	11
Maldah	35.62	17	50.71	17
Murshidabad	38.28	16	55.05	16
Birbhum	48.56	11	62.16	14
Burdwan	61.88	6	71.00	7
Nadia	52.53	9	66.55	10
North 24-Parganas	66.81	4	78.49	2
South 24-Parganas	55.10	8	70.16	8
Hooghly	66.78	5	75.59	4
Bankura	52.04	10	63.84	12
Purulia	43.29	15	56.14	15
Midnapore	69.32	3	75.17	5
Howrah	69.62	2	77.64	3
Kolkata	70.18	1	81.31	1
West Bengal	57.70		69.22	

Rank correlation coefficient: +0.95

g) Percentage of Main Workers in Total Population

The percentage of main workers in total population can lead an economy to the phase of prosperity. The higher this percentage, in a region, the more advanced will be that region and vice versa. Due to non-availability of this information in 2001 we have considered this aspect in the year 1981 and 1991.

The districts where the percentage of main workers in total population has been less than the state average figure have been considered to be in the state of depression. In 1981, there were nine such districts: Nadia, 24-Parganas, Murshidabad, Howrah, Birbhum, Midnapore, Hooghly, Maldah and Cooch Bihar. However, in 1991 there was a reduction in the number of such districts and these districts became 24-Parganas (both), Nadia, Howrah, Burdwan and Murshidabad. Thus in between these two years with a gap of ten years, Birbhum, Midnapore, Hooghly, Maldah and Cooch Bihar showed the sign of improvement.

Table 11.7: Percentage of Main Workers in the Total Population of the Districts of West Bengal and Ranks Attained from Maximum to Minimum

Districts	1981		1991	
	Percentage of Main Workers in Total Population	Rank attained	Percentage of Main Workers in Total Population	Rank attained
Darjeeling	36.46	2	34.22	2
Jalpaiguri	35.17	4	33.79	8
Cooch Bihar	29.85	11	32.15	9.5
North Dinajpur	32.17	8	34.28	3.5
South Dinajpur	32.17	8	34.86	6
Maldah	30.30	10	34.69	7
Murshidabad	28.76	13	31.51	13
Birbhum	30.76	9	33.14	12
Burdwan	32.72	6	30.66	14
Nadia	27.41	15	29.34	16
North 24-Parganas	28.13	14	28.78	17
South 24-Parganas	28.13	14	28.29	18
Hooghly	29.83	12	31.14	9.5
Bankura	32.90	5	35.91	3.5
Purulia	38.48	1	43.16	1
Midnapore	32.45	7	35.19	11
Howrah	27.19	16	29.10	15
Kolkata	35.25	3	33.01	5
West Bengal	30.98	-	32.19	-

Rank correlation coefficient: +0.81

All the eighteen districts of this State in both 1981 and 1991 have been ranked on the basis of this criterion and the rank correlation coefficient with regard to this variable among these districts in 1981 and 1991 has been found to be +0.81. This indicates that in respect of this variable there has been some minor change in the ranking of the districts in the years 1981 and 1991.

RESULTS OF OUR SEARCH

For determining whether a district can be considered depressed or not we have taken account of the earlier-stated seven criteria and the results of our search have precisely been presented in Tables 11.1 to 11.8.

Any district satisfying at least 4 (more than 50% of the criteria) has been identified depressed, as revealed by the record of the census years 1991 and 2001. Table 11.8 brings out that in 1991 as many as 12 districts responded to these criteria whereas in 2001, 11 districts did the same. The common districts satisfying these indicators in both the censuses are eleven in number. All the districts of North Bengal barring Darjeeling fall in this category. Besides *Murshidabad*, a neighbouring district of *North Bengal*, the districts of *Nadia, South 24-Parganas, Birbhum, Bankura* and *Purulia* revealed these features. These are, therefore, the most depressed districts of West Bengal.

The relatively advanced districts, according to our criteria, are:

1. *Hooghly*, 2. *Kolkata*, 3. *Midnapore*, 4. *Howrah*, 5. *North 24-Parganas* and 6. *Burdwan*. *Darjeeling*, a district of North Bengal and one of the attractive districts to the tourists, just lies on the border line.

There are many other indicators, say, growth rate of agriculture, growth rate of industry, growth rate of per capita income, per capita bank deposits and so on which, when applied on the districts of West Bengal, may eliminate some of the eleven districts from being depressed and at the same time, may bring some new districts in the region of depression. Our analysis does not go in this direction. Again, we have identified only the districts as most depressed. But within the districts there are so many areas which do not thrive in

Table 11.8: Depressed Districts of west Bengal

Criteria	Censuses years and districts		Remarks*
	1991	2001	
1. Growth Rate of Population (more than the State Average)	North Dinajpur, North 24-Parganas, South 24-Parganas, Nadia, Maldah, Murshidabad, Darjeeling, Howrah and Burdwan = 9	North Dinajpur, Maldah, Murshidabad, Darjeeling, North 24-Parganas, South Dinajpur, Jalpaiguri, South 24-Parganas, Nadia and Birbhum=10	-South Dinajpur, +Howrah, +Burdwan
2. Density of Population (less than the State Average)	Purulia, Bankura, Darjeeling, Jalpaiguri, South Dinajpur, Birbhum, South 24-Parganas, Midnapore, North Dinajpur, Cooch Behar and Maldah=11	Purulia, Bankura, Darjeeling, Jalpaiguri, South Dinajpur, Birbhum, South 24-Praganas, Midnapore, North Dinajpur, Cooch Behar and Maldah=11	0
3. Rate of Urbanization (less than the State Average)	Maldah, Cooch Behar, Bankura, Birbhum, Purulia, Midnapore, Murshidabad, North Dinajpur, South Dinajpur, Jalpaiguri and Nadia=11	Maldah, Cooch Behar, Bankura, Birbhum, Purulia, Midnapore, Murshidabad, North Dinajpur Jalpaiguri and Nadia=10	+South Dinajpur
4. Sex Ratio (less than the State Average	Kolkata, Howrah, Burdwan, North 24-Parganas and Darjeeling=5	Kolkata, Howrah, Burdwan, North 24-Parganas=4	-North 24-Parganas
5. Literacy Rate (less than the State Average)	North Dinajpur, Murshidabad, Maldah, Purulia, Jalpaiguri, Cooch Behar, South Dinajpur, Birbhum, Bankura, Nadia and South 24-Parganas=11	North Dinajpur, Murshidabad, Maldah, Purulia, Jalpaigruri, Cooch Behar, South Dinajpur, Birbhum, Bankura and Nadia=10	+South 24-Parganas

Contd. . .

Criteria	Censuses years and districts 1991	2001	Remarks*
6. Percentage of SC and ST people in total population of the district (more than the State average)	Jalpaiguri, Cooch Behar, South Dinajpur, Bankura, Purulia, Birbhum, South 24-Parganas, North Dinajpur, Burdwan, Nadia and Darjeeling=11	Jalpaiguri, Cooch Behar, South Dinajpur, Bankura, Purulia, Birbhum, South 24-Parganas, North Dinajpur, Burdwan, Nadia and Darjeeling=11	0
7. Percentage of main workers in total population of the district (less than the State average)	Howrah, Nadia, 24-Parganas, Murshidabad, Hooghly and Cooch Behar=6	South 24-Paragans, North 24-Parganas, Nadia, Howrah, Burdwan and Murshidabad=6	+ Hooghly, +Cooch Behar, -Burdwan

Summarised form	1991	2001
Depressed Districts**	Cooch Behar (5), North Dinajpur (5), Nadia (5), South 24-Parganas (5), Darjeeling (5), Jalpaiguri (4), South Dinajpur (4), Maldah (4), Murshidabad (4), Birbhum (4), Bankura (4), and Purulia (4)=12	Jalpaiguri (5), North Dinajpur (5), Birbhum (5), Nadia (5), Cooch Behar (4), South Dinajpur (4), Maldah, Murshidabad (4), South 24-Parganas (4), Bankura (4) and Purulia (4)=11

Most Depressed Districts

North Dinajpur (5), Nadia (5), Cooch Behar (4), South 24-Parganas (4), Jalpaiguri (4), South Dinajpur (4), Maldah (4), Murshidabad (4), Birbhum (4), Bankura (4) and Purulia (4)=11

*+indicates Improvement; -indicates deterioration; 0 indicates unchanged

**Figures within the parenthesis indicate the number of criteria of depression the district concerned upheld.

the same manner. This means that there are intra-district variations in the growth rate of different aspects of population. We have not proceeded in this direction also. For this further course of research is necessary.

The rank correlation coefficient in case of all the indicators among the eighteen districts of West Bengal in 1991 and 2001 (shown at the bottom of every table) has been seen to be almost perfect. This inspires us to conclude that with respect to the indicators of backwardness of a district considered here, the ranking of the districts of West Bengal has almost remained unchanged.

The surprising result derived from our search is thus the ranking of the districts with reference to the major demographic features has remained more or less unaltered in 2001 in comparison to what that had been in 1991.

CONCLUSION

The population data as thrown up by the last few censuses reveal that there has been a fall in the growth of population in the rural areas and a steep rise in it in the urban areas. The fact appears to be somewhat puzzling. But it is a reality. Is it then that the family planning programme is going to be successful in rural areas and a failure in urban areas? No, that is not so. Urbanization has been taking place at a rapid pace not due to development as such but due to the migration of the people from the rural areas to the urban one (that is, towns, cities and the outskirts of the cities) in search of service, shelter and better living. As a consequence, the urban areas are becoming congested, polluted, unhealthy and unbearable and thereby these areas are growing to be much more depressed. So unless the depression in rural areas is removed, the depression in urban areas will continue to crop up and the cumulative effect will be that the State as a whole will remain in a state of depression.

REFERENCES

1. Bhagwati, J. (1987): The Economics of Underdeveloped Counties, All India Traveller Book Seller, D-2/15, Krishan Nagar, Delhi: 110051.
2. Bose, Manjula (1984-85): "Rural Economy of West Bengal—an Inter-district Study", Artha Shastra, Vol. 3, No.s 1-2.

3. Census of India: Various Issues.

4. Economic Review, Government of West Bengal: Various Issues.

5. Gill, R. T. (1984): Economic Development: Past and Present, Prentice-Hall of India Private Limited, New Delhi 110001.

6. Konar, D. N. and Konar, B. N. (1989): "A Profile of India's Demographic Pattern Since 1981," *Economic Affairs*, Vol. 34, Qr. 4.

7. Konar, D. N. (1990): "A Search for Depressed Areas of West Bengal", *Artha Shastra*, Vol. 9, No. 1.

8. Konar, D. N. (2002): "The Inter-state Pattern of Human Development in India", *Artha Beekshan*, Vol. 11, No.3, December.

12

THE INTER-STATE PATTERN OF HUMAN DEVELOPMENT IN INDIA

In this chapter the basic objective is to bring into light the pattern of human development that has been experienced by the majority of States of our country. Over the last twenty years or so a significant emphasis has been put on various aspects of human development. The pioneers in this field are experts of the United Nations Development Programme (UNDP), mainly late Dr. Mahbub-ul-Haq and Professor Amartya Sen, the Nobel Laureate in Economics. Again, with the introduction of the concept of "human development index" the importance of the study of human development has increased to a great extent. With respect to the magnitude of population, India is the second largest country in the world and according to the census of 2001 we have a population figure of 1020 million. Everyone knows it. But what everyone does not know at all is various aspects of population of our country. Moreover, the inter-state pattern of human development is also not known to many. It is absolutely necessary to know this pattern attained by the major States. In reality, the true picture of an economy to the size of India can be ascertained only by studying the success and failures of our States in different directions. Then and then only we can prescribe a number of policies which, if followed, may bring more development even at the State level and that is also necessary for fostering quick growth and development at the macro level.

A SHORT PROFILE OF INDIA

India is one of the oldest civilizations with a Kaleidoscopic and rich cultural heritage. It has achieved multifarious socio-economic progress during the last fifty seven years of its independence. It measures 3214 km. from North to South and 2933 km. from East to West with a total land area of 3287263 sq. km. It has a land frontier of 15200 km. and a coastal line of 75165 km. Andaman and Nicobar Islands in the Bay of Bengal and Lakshadweep in the Arabian sea are parts of India. India, accounting for only 2.42 percent of the total global land, has given home to about 16 percent of the world population. Naturally, ours is an over-populated country. Hence it is imperative to know various aspects of population not only of India but also at least of major States. So a brief discussion about diverse nature of Indian States is very necessary.

DIVERSE NATURE OF INDIAN STATES

All the information depicted in the previous section is in the context of India as a whole. But India is a vast country where there are 28 States and 7 Union Territories—Chhattisgarh, Jharkhand and Uttaranchal are three new States. The average density of population per sq. km. is 324 (Census 2001) and it varies considerably from one State to another State. For illustration, it is as high as 904 in case of West Bengal and as low as only 13 in Arunachal Pradesh. It is the country principally of the Hindus as about 82 percent of our population belongs to this religion. Muslims (12.0.%), Christians (2.3%), Sikhs (2.0%), Jains (0.4%) and people of some other religions also live here. People of all these religions have made our India "Great". We are proud to say that two of our four metropolitan cities, namely Mumbai and Kolkata had respectively occupied the sixth and tenth place in the race of the twelve highly populated cities of the world. Ours is a diverse country. Looking at its diversity the Greek Emperor, Alexander the Great, while coming to conquer India, had once said to his

Army Chief, Selukas, "It's really true, Selukas... How wonderful is this land!" This statement is equally true even today, after 2325 years! Obviously, the nature, progress and degree of development among different States have been vividly divergent as the States themselves are virtually different among themselves in many a respect, namely area, physical appearance, population, culture, natural endowment and so on. Consequently, some States have grown at a rapid pace and have attained higher human development index while some others have progressed at a snail's pace and have virtually remained backward for years together.

HUMAN DEVELOPMENT AND ITS INDICATORS

Human development, though not a new concept (it had clearly been mentioned in the form of a song made prominent by a Bengalee saint "Ramprasad" who used to live in Halisahar (near Kalyani in West Bengal), has been made popular by the recent literature published by the United Nations Development Programme (UNDP) since 1990. This programme has emphasized: "Human development is the end- economic growth is the means. So the purpose of growth should be to enrich people's lives". However, it is not possible to give a unique definition of this term. Like economic development which is more pervasive than economic growth, human development can be specified and measured not by a particular indicator but by a series of indicators. In this paper we like to highlight as many as eighteen indicators: 1. birth rate, 2. death rate, 3. growth rate of population, 4.total fertility rate, 5. infant mortality rate , 6 maternal mortality rate, 7. density of population , 8. literary rate, 9. expectation of life at birth , 10. sex ratio, 11. urbanization, 12. percentage of people below poverty line, 13. per capita net state domestic product (PCNSDP), 14. anaemia among women, 15. anaemia among children, 16. working children in the age-group 5 to 14, 17. human poverty index and 18. human development index.

We intend to examine the nature and diversity of human development attained by the major States of our country in the light of the foregoing criteria. By major States we have meant those States whose population, according to 1991 census, was at least 10 million. These States in the alphabetical order of their names are:

1. Andhra Pradesh (A.P.), 2. Assam, 3. Bihar, 4. Gujarat, 5. Haryana, 6. Karnataka, 7. Kerala, 8.Madhya Pradesh (M.P)., 9. Maharashtra, 10. Orissa, 11. Punjab, 12. Rajastahn, 13. Tamil Nadu, 14. Uttar Pradesh (U.P) and 15. West Bengal. (W.B).

METHODOLOGY OF THE STUDY

Our study has been an empirical one being based on official data which have been collected from different Government publications. These data have been compiled and arranged to suit our purpose. For a variable like growth rate of population we have computed the annual compound growth rate by simply fitting the exponential curve $P_t=ab^t$ by the usual procedure and the annual compound growth rate has, obviously, been r =(b-1) x100% . The information on human poverty index (HPI) and human development index (HDI) have been taken from National Human Development Report, 2002. All the fifteen States in respect of each of eighteen indicators have been identified by giving ranks in the usual manner on the basis of their performance.

The States concerned have been grouped into three categories, namely A–highly developed, B–moderately developed and C–Undeveloped. The desiderata for being included in each of those three categories by the States concerned with respect to each of eighteen indicators have been quantified in column 1 of Tables 12.4, 12.5 and 12.6. On the basis of the overall analysis any State satisfying at least ten (that is , more than 50 percent of total criteria) such criteria will be said to have a place in the relevant category.

The identity of two or more States holding the same number of criteria of falling in a particular category (say, A) will be settled by looking into the priority of actual ranks attained by the States in that category. The actual position of two or more States satisfying the same number of criteria of falling in category B (or C) will be fixed up by examining not only the number of criteria held by them in that category B (or C) but also by examining one state's supremacy over the other in actual rank in that category together with the examination of the same in category A and C and so on.

RESULTS OF THE STUDY

The results of the present study have primarily been put in Tables 12.1, 12.2 and 12.3 signifying the above-named indicators of human development and attainment of the States to those indicators and the appropriate ranks in such attainment put within the parenthesis. Table 12.4 represents categorization of States on the basis of their attainment to those indicators while Table 12.5 brings into light the number of criteria the States satisfy for being considered in either of the three categories. Table 12.5 has, primarily, become the decisive factor in determining the relative position of the States in the ultimate analysis.

CONCLUSION

The foregoing discussion and the picture meticulously presented in seven tables, particularly the last one (Table 12.5), enable us to automatically make the following conclusions:

1. Kerala, a Southern State, is at the topmost position among the fifteen major States we have brought in this paper. This is because of the fact that this State has occupied the top rank in as many as thirteen indicators, third rank in case of two indicators, seventh, ninth and tenth rank in case of one indicator each. In our judgment this State has satisfied fifteen out of eighteen criteria (that is, 83 percent) for being considered as a State belonging to A –Category. It may also be noted that another State of South, namely Tamil Nadu, has fulfilled the same number of criteria (that is, fifteen) for being included as a State of A-type. But a scrutiny of the relative performance of these two States of South will simply put Kerala far ahead of Tamil Nadu. Hence, Tamil Nadu, in our opinion, has occupied the second place in A-Category. Maharashtra and Punjab can jointly claim to be a State possessing the third place among States of A-Category. But Punjab has the edge on Maharashtra in that this State has got second rank in case of five indicators while Maharashtra has done so in only three cases. The

Table 12.1: Indicators of Human Development for Major States

Sl.No.	States	Birth Rate (per 1000) 1999	Death Rate (per 1000) 1999	Growth Rate of Population p.a. % (1991-2001)	Total Fertility Rate (No. of Children) (1995-1997)	Infant Mortality Rate (Per 1000 Live Births) 1999	Maternal Mortality Rate (per 100000 Live Births) 1998
1.	Andhra Pradesh	21.7 (6)	8.2 (9)	1.31 (3)	2.8 (5)	66 (9)	159 (5)
2.	Assam	27.0 (11)	9.7 (12)	1.74 (7)	3.3 (10)	76 (11)	409 (11)
3.	Bihar	31.5 (14)	8.9 (11)	2.43 (13)	4.5 (14)	63 (7)	452 (12)
4.	Gujarat	25.4 (9)	7.9 (7)	2.05 (9)	3.1 (8)	63 (7)	28 (1)
5.	Haryana	26.8 (10)	7.7 (5)	2.50 (14)	3.5 (14)	68 (10)	103 (3)
6.	Karnataka	22.3 (7)	7.7 (5)	1.60 (5)	2.6 (3)	58 (6)	195 (6)
7.	Kerala	18.0 (1)	6.4 (1)	0.90 (1)	1.8 (1)	14 (1)	198 (7)
8.	Madhya Pradesh	31.1 (12)	10.4 (13)	2.06 (10)	4.1 (12)	90 (14)	498 (13)
9.	Maharashtra	21.1 (4)	7.5 (4)	2.06 (10)	2.8 (5)	48 (2)	135 (4)
10.	Orissa	24.1 (8)	10.7 (15)	1.49 (4)	3.1 (8)	97 (15)	367 (10)
11.	Punjab	21.5 (5)	7.4 (3)	1.82 (8)	2.8 (5)	53 (5)	199 (8)
12.	Rajasthan	31.1 (12)	8.4 (10)	2.52 (15)	4.2 (13)	81 (12)	670 (14)
13.	Tamil Nadu	19.3 (2)	8.0 (8)	1.07 (2)	2.1 (2)	52 (3)	79 (2)
14.	Uttar Pradesh	32.8 (15)	10.5 (14)	2.29 (12)	4.9 (15)	84 (13)	707 (15)
15.	West Bengal	20.7 (3)	7.1 (2)	1.65 (6)	2.7 (4)	52 (3)	266 (9)
	All India	26.1	8.7	1.95 3.4	70	407	

Sources:Arranged from data taken from

a) Office of the Registrar General, India,

b) Economic Survey, Government of India 2001-2002,

c) National Human Development Report,2001.

Note: Figures within parenthis indicate usual ranks.

Table 12.2: Indicators of Human Development for Major States

Sl.No.	States	Density of population (2001)	Literacy Rate (2001)	Life Expectancy at Birth (1993-1997)	Sex Ratio (2001)	Urban population as percentage of total population (2001)	Percentage of people below poverty line (1999-2000)
1.	Andhra Pradesh	689 (4)	61.1 (12)	62.4 (8)	978 (3)	27.1 (8)	15.8 (6)
2.	Assam	340 (9)	64.3 (9)	56.7 (14)	932 (7)	12.7 (14)	36.1(12)
3.	Bihar	880 (2)	47.5 (15)	59.6 (11)	921(10)	10.5 (15)	42.6(14)
4.	Gujarat	258 (12)	66.4 (8)	61.9 (9)	921(10)	37.3 (3)	14.1 (4)
5.	Haryana	477 (8)	68.6 (6)	64.1 (4)	861(15)	29.0 (6)	8.7 (2)
6.	Karnataka	275 (11)	67.0 (7)	63.3 (6)	964 (5)	34.0 (4)	20.0 (7)
7.	Kerala	819 (3)	90.9 (1)	73.3 (1)	1058 (1)	26.0 (10)	12.7 (3)
8.	Madhya Pradesh	196 (14)	64.1 (10)	55.5 (15)	920(12)	26.7 (9)	37.4(13)
9.	Maharashtra	314 (10)	77.3 (2)	65.5 (3)	922 (8)	42.4 (2)	25.0 (9)
10.	Orissa	236 (13)	63.6 (11)	57.2 (13)	972 (4)	15.0 (13)	47.2(15)
11.	Punjab	482 (6)	69.9 (4)	67.7 (2)	874(14)	33.9 (5)	6.2 (1)
12.	Rajasthan	165 (15)	61.0 (13)	60.0 (10)	922 (8)	23.4 (11)	15.3 (5)
13.	Tamil Nadu	478 (7)	73.5 (3)	64.1 (4)	986(2)	43.9 (1)	21.1 (8)
14.	Uttar Pradesh	689 (4)	57.4 (14)	57.6 (12)	898(13)	20.8 (12)	31.2(11)
15.	West Bengal	904 (1)	69.2 (5)	62.8 (7)	934 (6)	28.0 (7)	27.0(10)
	All India	324	65.2	61.1	933	27.7	26.1

Sources: Arranged from data taken from
a) Office of the Registrar General, India,
b) Census of India 2001, Series 20, West Bengal,
c) National Human Development Report,2001,
and d). Economic Survey, Govt. of India 2001-02.
Figures within parenthesis indicate usual ranks.

Table 12.3: Indicators of Human Development for Major States

Sl.No.	States	Per capita net state domestic product (Rs.at 1980-81 prices), 1997-98	Anaemia among women (in percentage) 1998-99.	Anaemia among children (in percentage), 1998-99.	Working children in the age–group 5 to 14 (in percentage), 1991.	Human Poverty Index, 1991.	Human Development Index, 1991.
1.	Andhra Pradesh	2550 (8)	49.8 (9)	72.3 (5)	10.0 (15)	39.78 (8)	0.344 (9)
2.	Assam	1675 (13)	69.7(15)	63.2 (2)	5.5 (9)	48.95 (13)	0.326 (11)
3.	Bihar	1126 (15)	63.4(14)	81.3 (13)	4.0 (5)	52.34 (15)	0.286 (13)
4.	Gujarat	3918 (4)	46.3 (4)	74.5 (8)	5.3 (8)	29.46 (6)	0.380 (6)
5.	Haryana	4025 (3)	47.0 (5)	83.9 (15)	2.6 (2)	28.55 (3)	0.409 (4)
6.	Karnataka	2866 (7)	42.4 (3)	70.6 (4)	8.8 (14)	32.70 (7)	0.367 (8)
7.	Kerala	2490 (9)	22.7 (1)	43.9 (1)	0.6 (1)	19.93 (1)	0.576 (1)
8.	Madhya Pradesh	1922 (11)	54.3(10)	75.0 (9)	8.1 (13)	43.47 (10)	0.282 (15)
9.	Maharashtra	5032 (1)	48.5 (6)	76.0 (10)	5.7 (10)	29.25 (4)	0.403 (5)
10.	Orissa	1666 (14)	63.0(13)	72.3 (5)	5.9 (11)	49.85 (14)	0.328 (10)
11.	Punjab	4389 (2)	41.4 (2)	80.0 (12)	3.0 (3)	25.06 (2)	0.447 (2)
12.	Rajasthan	2226 (10)	48.5 (6)	82.3 (14)	6.5 (12)	46.67 (11)	0.298 (12)
13.	Tamil Nadu	3141 (5)	56.5(11)	69.0 (3)	4.8 (7)	29.28 (5)	0.421 (3)
14.	Uttar Pradesh	1725 (12)	48.7 (8)	73.9 (7)	3.8 (4)	48.27 (12)	0.284 (14)
15.	West Bengal	2977 (6)	62.7(12)	78.3 (11)	4.2 (6)	40.48 (9)	0.370 (7)
	All India	2840	51.8	74.3	5.4	39.36	0.340

Sources: Arranged from data taken from

a) Office of the Registrar General, India,

b) National Human Development Report,2001.

and c) Economic Survey, Govt. of India 2001-02.

Figures within parenthesis indicate usual ranks.

Table 12.4: Categorisation of States of India on the Basis of their Attainment in Human Development

INDICATORS OF HUMAN DEVELOPMENT	A- CATEGORY	B-CATEGORY	C- CATEGORY
1. BIRTH RATE : A. Upto 22.0 B. Between 22.0 and 30.0 C. Above 30.0	1.Kerala, 2. Tamil Nadu, 3.West Bengal, 4. Maharashtra , 5. Punjab and 6. Andhra Pradesh.	7.Karnataka, 8. Orissa, 9. Gujarat, 10. Haryana and 11.Assam.	12. Madhya Pradesh, 13. Rajashthan, 14. Bihar and 15. Uttar Pradesh.
2. DEATH RATE : A. Upto 8 B. Between 8 and 10 C. Above 10.	1.Kerala, 2.West Bengal, 3.Punjab, 4. Maharashtra , 5.Karnataka, 6.Haryana, 7.Gujarat and 8. Tamil Nadu.	9. Andhra Pradesh, 10.Rajasthan, 11.Bihar and 12. Assam.	13. Madhya Pradesh , 14. Uttar Pradesh and 15. Orissa.
3. GROWTH RATE OF POPULATION: A. Upto 1.75% B. Between 1.75% and 2.25% C. Above 2.25%	1. Kerala, 2.Tamil Nadu, 3. Andhra Pradesh, 4.Orissa, 5. Karnataka, 6.West Bengal and 7. Assam.	8. Punjab, 9. Gujarat, 10.Maharashtra and 11. Madhya Pradesh.	12. Uttar Pradesh, 13. Bihar, 14. Haryana and 15. Rajasthan.
4. TOTAL FERTILITY RATE : A. Upto 3 B. Between 3 and 4 C. Above 4.	1. Kerala, 2.Tamil Nadu, 3. Karnataka, 4.West Bengal, 5. Punjab, 6. Maharashtra and 7. Andhra Pradesh.	8. Orissa, 9. Gujarat, 10.Assam and 11. Haryana.	12. Madhya Pradesh, 13. Rajasthan, 14. Bihar and 15. Uttar Pradesh .
5. INFANT MORTALITY RATE : A. Upto 60 B. Between 60 and 75 C. Above 75.	1. Kerala, 2.Maharashtra, 3.Tamil Nadu, 4. West Bengal, 5. Punjab and 6. Karnataka.	7.Bihar, 8. Gujarat, 9. Andhra Pradesh and 10.Haryana.	11. Assam, 12. Rajasthan, 13. Uttar Pradesh, 14. Madhya Pradesh and 15. Orissa.
6. MATERNAL MORTALITY RATE : A. Upto 175 B. Between175 and 400.	1. Gujarat, 2.Tamil Nadu, 3.Haryana, 4. Maharashtra, 5. Andhra Pradesh.	6. Karnataka, 7. Kerala, 8. Punjab, 9. West Bengal and 10. Orissa.	11. Assam, 12. Bihar, 13. Madhya Pradesh, 14. Rajasthan and

contd. . .

C. Above 400.			15. Uttar Pradesh.
7. DENSITY OF POPULATION : A . Above 600 B. Between 400 and 600 C. Less than 400.	1.West Bengal, 2.Bihar, 3. Kerala, 4. Andhra Pradesh and 5. Uttar Pradesh.	6.Punjab, 7.Tamil Nadu and 8. Haryana.	9.Assam, 10. Maharashtra, 11.Karnataka, 12. Gujarat, 13. Orissa, 14. Madhya Pradesh and 15. Rajasthan.
8. LITERACY RATE: A. Above 65% B. Between 60% and 65%. C. Less than 60%.	1. Kerala, 2.Maharashtra, 3.Tamil Nadu,4.Punjab, 5. West Bengal. 6.Haryana , 7. Karnataka and 8. Gujarat.	9.Assam, 10.Madhya Pradesh, 11.Orissa, 12.Andhra Pradesh and 13. Rajasthan.	14.Uttar Pradesh and 15.Bihar.
9. LIFE EXPECTANCY AT BIRTH : A. Above 64 years. B. Between 60 and 64 years. C. Less than 60 years.	1.Kerala, 2.Punjab, 3.Maharashtra, 4.Tamil Nadu and 5. Haryana.	6. Karnataka, 7.West Bengal, 8. Andhra Pradesh, 9. Gujarat and 10. Rajasthan.	11.Bihar, 12. Uttar Pradesh, 13. Orissa, 14. Assam and 15. Madhya Pradesh.
10. SEX RATIO : A. 950 and over, B. Between 900 and 950 C. Below 900.	1. Kerala, 2. Tamil Nadu , 3. Andhra Pradesh, 4.Orissa and 5. Karnataka.	6. West Bengal, 7.Assam, 8. Maharashtra, 9. Rajasthan, 10. Bihar, 11. Gujarat, 12. Madhya Pradesh.	13. Uttar Pradesh, 14. Punjab and 15. Haryana.
11.PERCENTAGE OF URBAN POPULATION IN TOTAL POPULATION : A. Above 30%. B. Between 25% and 30%. C. Below 25%.	1.Tamil Nadu, 2. Maharashtra, 3.Gujarat, 4.Karnataka and 5. Punjab.	6. Haryana, 7.West Bengal, 8.Andhra Pradesh, 9.Madhya Pradesh and 10. Kerala.	11. Rajasthan, 12. Uttar Pradesh, 13.Orissa, 14.Assam and 15. Bihar.
12.PERCENTAGE OF PEOPLE BELOW POVERTY LINE: A. Below 20%.	1. Punjab,2.Haryana, 3.Kerala, 4. Gujarat,5. Rajasthan and 6. Andhra Pradesh.	7. Karnataka, 8. Tamil Nadu, 9. Maharashtra and 10. West Bengal.	11. Uttar Pradesh, 12.Assam, 13.Madhya Pradesh,

Indicator	A	B	C
B. Between 20% and 30%. C. Above 30%.			14.Bihar and 15. Orissa.
13.PER CAPITA NSDP : A : Above Rs.3000 B : Between Rs.2000 to Rs.3000 C : Below Rs.2000.	1. Maharashtra, 2. Punjab, 2. Haryana, 4. Gujarat and 5. Tamil Nadu.	6. West Bengal, 7.Karnataka, 8. Andhra Pradesh, 9. Kerala and 10. Rajasthan.	11. Madhya Pradesh, 12.Uttar Pradesh, 13.Assam, 14. Orissa and 15.Bihar.
14.ANAEMIA AMONG WOMEN: A : Upto 50 percent B : Between 50 and 60 percent C : Above 60 percent.	1. Kerala, 2. Punjab, 3. Karnataka, 4.Gujarat, 5. Haryana, 6.Maharashtra, 7. Rajasthan, 8. Uttar Pradesh and 9. Andhra Pradesh.	11. Madhya Pradesh and 12. Tamil Nadu.	12.West Bengal, 13. Orissa, 14. Bihar and 15. Assam.
15. ANAEMIA AMONG CHILDREN: A: Upto 75 percent B : Between 75 and 80 percent C : Above 80 percent.	1. Kerala, 2.Assam, 3.Tamil Nadu, 4. Karnataka, 5. Orissa, 6. Andharapradesh, 7. Uttar Pradesh, 8. Gujarat and 9. Madhya Pradesh.	10.Maharashtra,11.West Bengal and 12. Punjab.	13. Bihar, 14. Rajasthan and 15. Haryana.
16. WORKING CHILDREN: A : Upto 5 percent B : Between 5 and 7 percent C : Above 7 percent.	1. Kerala, 2. Haryana, 3. Punjab, 4. Uttar Pradesh, 5. Bihar, 6. West Bengal and 7. Tamil Nadu.	8. Gujarat, 9. Assam, 10. Maharashtra, 11. Orissa and 12. Rajasthan.	13. Madhya Pradesh, 14. Karnataka and 15. Andhra Pradesh .
17. HUMAN POVERTY INDEX: A: Upto 30.00 B: Between 30.00 and 45.00 C: Above 45.00	1. Kerala, 2. Punjab, 3. Haryana, 4. Maharastra, 5. Tamil Nadu and 6. Gujarat.	7. Karnataka, 8. Andhra Pradesh, 9. West Bengal. and 10. Madhya Pradesh.	11. Rajasthan, 12. Uttar Pradesh, 13. Assam, 14. Orissa and 15. Bihar.
18. HUMAN DEVELOPMENT INDEX: A : Above 0.400 B : Between 0.300 and 0.400 C : Below 0.300	1. Kerala, 2. Punjab, 3. Tamil Nadu, 4. Haryana and 5. Maharashtra.	6. Gujarat, 7. West Bengal, 8. Karnataka, 9.Andhra Pradesh, 10. Orissa and 11. Assam.	12. Rajasthan, 13. Bihar, 14. Uttar Pradesh and 15. Madhya Pradesh.

fifth place in the race for States belonging to A-Category goes to Haryana which is second only to Punjab so far as the State's capacity for reduction of poverty is concerned and is also second only to Kerala with respect to the percentage of working children in the age-group 5 to 14. No other State, in our opinion, can claim to be a State belonging to A- category.

Table 12.5: Major States and the Number of Criteria Fulfilled for being considered as a State of Various Categories

Sl.No.	States	A-Category	B–Category	C-Category	Total
1.	Andhra Pradesh	9	8	1	18
2.	Assam	2	7	9	18
3.	Bihar	2	3	13	18
4.	Gujarat	9	8	1	18
5.	Haryana	10	5	3	18
6.	Karnataka	8	8	2	18
7.	Kerala	15	3	0	18
8.	Madhya Pradesh	1	6	11	18
9.	Maharashtra	13	4	1	18
10.	Orissa	3	6	9	18
11.	Punjab	13	4	1	18
12.	Rajasthan	2	6	10	18
13.	Tamil Nadu	15	3	0	18
14.	Uttar Pradesh	4	0	14	18
15.	West Bengal	8	9	1	18

Source: Summarised from Table 12.4.

2. Both Andhra Pradesh and Gujarat may be considered as topmost States of B-Category as both of them have fulfilled nine criteria of being a successful State in category A and eight criteria of being included as a State of B-Category. However, Gujarat is a bit superior to Andhra Pradesh as in maternal mortality rate the former has headed all the States and some other aspects of human development have gone well with Andhra Pradesh. Judged by this consideration

Gujarat is the topper among States belonging to B – category while Andhra Pradesh just follows the former. There may be a tussle between Karnataka and West Bengal for finding a berth in the third place among States of B-category as both the States have held eight criteria of being a State of A-category. But a minute analysis of the first three tables will discern that West Bengal has an advantageous position over Karnataka simply because of the fact that with respect to density of population the State of West Bengal has topped the list of fifteen States and in respect of some other aspects of human development (say, death and birth rates) the performance of this State is better than that of Karnataka. Hence the third place among States of B-Category goes to West Bengal while the fourth one goes to Karnataka. The fifth place goes to Assam as this State on all-India basis has occupied the second rank in the indicator of anaemia among children and also the seventh rank in the growth rate of population, besides fulfilling seven criteria of being qualified as a State of B-Category.

3. In-fine, we may say that the remaining five States fall in C-category. Of them the position of Orissa, the land of God Jagannath, the land of Ashoke, the Great and the land of the famous temple of Konark, is the best as it has, on all-India basis, held the fourth rank in case of growth rate of population and sex ratio and fifth rank in anaemia among children. It has also possessed the eighth rank in birth rate and in total fertility rates. Next comes Rajasthan famous for both desert and historical places. The positions of Madhya Pradesh, Uttar Pradesh and Bihar are almost identical.

Thus the fifteen States arranged on the basis of our criteria of human development in diminishing order of importance become:

1. Kerala, 2. Tamil Nadu, 3. Punjab, 4. Maharashtra, 5. Haryana, 6. Gujarat, 7. Andhra Pradesh, 8. West Bengal, 9. Karnataka, 10. Assam, 11. Orissa, 12. Rajasthan, 13. Madhya Pradesh, 14. Uttar Pradesh and 15. Bihar.

Let us now find if there is any sort of relationship between level of human development attained by our States and level of economic growth as represented by per capita net State domestic product. This has been found out by calculating Spearman's rank correlation for which necessary data have been put in Table 12.6.

Table 12.6: Ranks of Indian States in the Attainment of Human Development (H) and Per Capita Net State domestic Product (P).

States	Rank in H	Rank in P
Kerala	1	9
Tamil Nadu	2	5
Punjab	3	2
Maharashtra	4	1
Haryana	5	3
Gujarat	6	4
Andhra Pradesh	7	8
West Bengal	8	6
Karnataka	9	7
Assam	10	13
Orissa	11	14
Rajasthan	12	10
Madhya Pradesh	13	11
Uttar Pradesh	14	12
Bihar	15	15

The rank correlation coefficient between H and P has been calculated to be 0.77. It may be noted that the same coefficient between Human Development Index attained by the States and per capita net State domestic product (that is, between column (7) and column (2) of Table 12.3) has been found to be 0.76. Hence we may say that there is little difference between our arrangement of States on the basis of human development and that made on the basis of human development index.

The above discussions, obviously, portray a scenario of striking contrasts and wide variations among major States of India. Some such contrasts will be as follows:

1. Higher human development with relatively lower income as made prominent by Kerala which heads all the States and Union Territories for the attainment of human development.
2. Lower human development at higher income as displayed by Maharashtra and Haryana.
3. Mutually re-inforcing growth and human development as revealed by Punjab, Gujarat, Tamil Nadu, Andhra Pradesh, West Bengal and Karnataka.
4. Mutually depressing growth and human development as made clear by Assam, Orissa, Rajasthan, Madhya Pradesh, Uttar Pradesh and Bihar.

The inter-State pattern of human development in India is, thus, full of many striking contrasts and there are dangers inherent in several patterns. States with lopsided pattern of development tilted against human development would be very soon in a dead end. The States with faster economic growth would see a trend of slowing down unless they undertake a massive programme of improving human development. Similarly, the topmost State in the sphere of human development, that is, Kerala should undertake programmes of accelerating economic growth to keep pace with human development so that it can reap the benefits of better human development by showing higher productivity.

Finally, the laggards like Assam, Orissa, Rajasthan, Madhya Pradesh, Uttar Pradesh and Bihar have been caught in the vicious circle of low human development and low economic growth. To get rid of this circle, these States are badly in need of a big push by fostering investment either first to push up growth and subsequently to pay attention to human development or to step up human development which would later put pressure to accelerate economic growth.

REFERENCES

1. Census of India, 2001, Series 20, West Bengal.
2. Census of India, Various Issues.

3. Census of India, 2001, Series 1, India, Provisional Population Tables.

4. Coale, A.J. and Hoover, E.M: Population Growth and Economic Development in Low-Income Countries.

5. Chatterjee, Pranab Kumar (1986) : "Human Aspect Development – Conceptual Empirical Study of Selected Asian Countries", *Arthashastra*, Vol. 5, No. 2.

6. Dreze, Jean and Sen, Amartya Kumar (1995): *Indian Economic Development and Social Opportunity*, Oxford University Press.

7. Datta, Bhabatosh: Economics of Industrialisation.

8. Datt, Ruddar and Sundharam, K.P.M (2002). *Indian Economy*, S. Chand & Company Ltd., New Delhi.

9. Datt, Ruddar and Konar, Dhirendranath (2000) : *Social Choice and Development*, Deep & Deep Publications Pvt. Ltd.

10. Government of India, *Economic Survey* (2000-01).

11. Higgins : *Economic Development*.

12. *Human Development Report* (2002) : Oxford University Press.

13. Konar, D.N. (1999) : "Pattern of Development of the Majority of States of India", *Economic Affairs*, Vol.44, Qr. 4. Dec.

14. Konar, D.N. (2001) : "Census 2001 and Sex Ratio in India with Special Reference to West Bengal" *Artha Beekshan* , Vol.10, No. 2, Dec.

15. Krishnaji, N (2000) : "Trends of Sex Ratio" *Economic and Political Weekly*, April 1-7.

16. Mazumdar, K & Paul, M (1987) : "Inter-state Variations in Human Resources and Economic Development", *Arthashastra*, Vol. 6, No. 2.

17. *National Human Development Report*, 2001.

18. Office of the Registrar General and Ministry of Health and Family Welfare, Annual Report (2000-01).

19. UNDP, *Human Development Report*, 1996 and 2000.

20. Visaria, P. and Visaria, L.: "India's Population Scene after 1981 Census" *Economic and Political Weekly*, Vol. XVI, No.s 44, 45 and 46, Special Number.

21. *World Development Report*, World Bank : 1990, 1997, 1998-99, 1999-2000 and 2000-2001.

13

HUMAN POVERTY IN INDIA AND OTHER SAARC COUNTRIES

INTRODUCTION

Poverty is a universal phenomenon and the concept of poverty is as old as the human civilization. It is a relative term: even the richest person of a region may think that he (she) is poorer than his (her) counterpart of another region. Hence to raise his (her) standard he (she) strives to earn more and more and this leads to more inequality in the economy. It is admitted that the famous **Pareto Law** introduced by **V. Pareto** in 1897 can, clearly, explain the inequality of size distribution (and even human capability). But earning of income which is the basic reason behind the existence of high inequality in a society, is more a matter of opportunity than a matter of human capability and cannot, therefore, be predicted by the said law.

By poverty we generally mean economic poverty caused due to the lack of purchasing power. It is widely believed that the manifestation of poverty is multi-dimensional in nature. Hence the estimates of poverty should be based not only by the income measure but also by a number of factors that may be called indicators of poverty. One such indicator is human poverty that is, again, due to high economic poverty.

SAARC AND ITS ESTABLISHMENT

Economic co-operation among a number of countries having some common characteristics has become one of the dominant features in the post second World War period. Numerous regional economies' alliances have taken shape after the Great War. One such alliance in Asia has been the establishment of South Asian Association for Regional Co-operation or the SAARC at a regional summit meeting in Dacca in the month of December 1985. The establishment of SAARC has been an important landmark in the history of interregional relations. Seven countries of South Asia, namely Bangladesh, Bhutan, India, Maldives, Nepal, Pakistan and Sri Lanka are the members of this regional co-operation. In the charter of SAARC adopted at the Dacca Summit it had been stated that SAARC is "desirous of promoting peace, stability, amity and progress in the region through strict adherence to the principle of non-use of force and peaceful settlement of disputes." It also calls for "fostering mutual understanding, good neighbouring relations and meaningful co-operation in the region."

Nature of the SAARC countries:

The countries forming the SAARC have some areas of commonality. Some such areas are: high pressure of population, low per capita income, low saving and investment ratio, much dependence on agriculture and on exports of primary commo-dity, wide-spread prevalence of poverty and dependence on foreign resources for development.

SAARC countries present a picture of asymmetry and economic scenario prevailing in these countries is not homogeneous. In one extreme there is a very big country, India and on the other, there is the smallest country, Maldives. In between them there are the remaining five countries revealing diversity of facts and figures.

Magnitude of poverty in the SAARC region:

According to the World Development Report 1985 about 29.7% of the global population lived in South Asia and of these people about 46.9% live on abject poverty. Poverty which is primarily a rural phenomenon, is extensive in India, Bangladesh, Pakistan, Nepal and Sri Lanka. The magnitude of poverty in these countries can be understood from Table 13.1.

Estimates of poverty are not available for Bhutan and Maldives; naturally, their cases have not been considered.

Table 13.1: Percentage of People Living Below the Poverty Line in the SAARC Countries

Countries	National poverty (%) in 1994		
	Rural (a)	Urban (b)	Total (c)
India	36.7	30.5	35.0
Bangladesh	39.8	14.3	35.6
Pakistan	36.9	28.0	34.0
Nepal	44.0	23.0	42.0
Sri Lanka	-	-	25.0

Source: World Development Report, 2003

It is seen from Table 13.1 that poverty is deep and extensive in the rural areas. It is also expected that the incidence of poverty will be higher in rural areas than in towns and cities. Coming to Indian context we can say that according to the census of 1981, 76.3% of our total population lived in the rural areas and they contributed only about 37% of the total Gross Domestic Product (GDP) whereas about 24% of our total population living in urban areas contributed about 63% of the GDP. What a contrast! So it is rightly said that poverty is a rural phenomenon. In fact, a large proportion of urban poverty is a spill over from villages. The poor people mostly consist of landless labourers, small and marginal farmers, rural artisans including fisherman and persons coming from backward classes and backward tribes.

It is known from the Development Report of the United Nations that today in the world the largest number of poor people live in India and the number of poor people in India is 233 million—this is about 22.5% of total population of India.

Human Poverty and its Indicators

Manifestation of poverty is multi-dimensional in nature. Hence the estimates of poverty should be based not only by the income measure but also by a number of factors called the indicators of poverty. One such indicator is human poverty whose basic symptoms are:

a) General trend of growth rate of population is high and density of population is also very high.

b) Condition of children with respect to health and education is very precarious.

c) Status of women with respect to sex ratio, maternal mortality rate and total fertility rate is highly miserable.

d) Adult illiteracy rate is very high.

e) Safe drinking water is inaccessible to a high percentage of people.

f) Health services are not within the reach of a high percentage of people.

g) A high percentage of people are deprived of the advantages of sanitary system.

h) Human development index is very low.

A short profile of the SAARC countries on the basis of the above indicators

A brief description of the countries forming the SAARC on the basis of the above indicators can now be made. In doing so we should, first of all, concentrate our attention on the population of the countries and the children of such countries. This is because the size of population of a country, how it has been growing and is likely to grow in the future, are some of the basic facts which largely influence every sector of economic and social development of a country. It is rightly said that the child is the father of a nation and today's children are tomorrow's leaders of the nation. Hence, the future of a nation depends very much upon health and proper education of the children of that nation. It is because of these reasons that we have given much emphasis upon the growth of population and the matters relating to children of the countries concerned in dealing with poverty.

A. Growth of population and density of population

Table 13.2 presents mátters relating to population and its growth rate, density of population and the corresponding percentage rate of change.

Table 13.2: Population and Density of Population of the SAARC Countries

Countries	1991		2001		Annual growth rate of population (%), 1991--2001.	Rate of change of density (%), 1991--2001.
	Population (million)	Density (No.)	Population (million)	Density (No.)		
India	862.7	264	1033.4	348	1.82	31.8
Bangladesh	116.4	768	133.4	1025	1.37	33.5
Pakistan	121.5	145	141.5	183	1.54	26.2
Nepal	20.1	138	23.6	165	1.62	19.6
Sri Lanka	17.4	261	19.6	304	1.20	16.5

Source: World Development Report, Various Issues.

It is clear from Table 13.2 that India has the maximum number of population. In fact, India is the second country in the world after China having maximum population. It is also seen from this Table that the annual compound growth rate of population of the SAARC countries is very high–the highest rate has, however, come from India followed by Nepal, Pakistan and Bangladesh. The growth rate of population has been the least in case of Sri Lanka. Bangladesh has the highest density of population and in this respect the position of India is the second. Naturally, the percentage rate of change of density of population has been maximum in case of Bangladesh followed by India.

B. Condition of children

We may now portray the condition of children in the SAARC countries as demonstrated through Table 13.3.

Table 13.3: Children of the SAARC Countries

Countries	Malnourished Children under age 5 in 1990 (in million)	Underweight children under age 5 in 1990 (in percentage)	Infant mortality rate (per 1000 live births) in 1990	Children not taking part in schools in 1990 (in million)
India	45.5	53	94	73.1
Bangladesh	12.6	56	114	21.2
Pakistan	12.0	38	104	25.7
Nepal	2.1	47	121	2.2
Sri Lanka	0.7	34	26	0.5

Source: World Development Report, Various Issues.

Table 13.3 authenticates the precarious condition of the children of the SAARC countries. A clear symptom of poverty gets reflected through malnourishment of children. At the same time it is poverty alone that deprives a large number of children of taking the basic education. The number of malnourished children and the number of them not taking part in primary and secondary schools, can be known from the above Table. The percentage of malnourished children under age 5 to the total population is 10.8, 10.5 and 9.9 respectively in Bangladesh, Nepal and Pakistan. In Sri Lanka this is 4.0 while in India this is 5.3. With such a high percentage of malnourished children what can be expected? It simply reminds us of the vicious circle of poverty–poverty is the cause and is itself the effect.

A significant symptom of widespread poverty lies with the higher infant mortality rate, that is, the annual number of deaths of infants under one year of age per 1000 live births. This rate is very high (121) in Nepal and (114) in Bangladesh. The corresponding rate is also high in Pakistan (104) and India (94). From this point of view Sri Lanka is in the supreme position. Another symptom of abject poverty in most of the countries of the SAARC is indicated by the fact that a large number of children are not simply taking part in primary and secondary schools. Table 13.3 also portrays the absolute number of such children. In Pakistan the number of children comprising of 21.2% of population is not taking basic education while in Bangladesh this percentage is 18.2. In Sri Lanka this percentage is the least (2.9) while in India this is 8.5.

C. Poor status of women in the society

A very important indicator of miserable poverty is poor status of women in the society. This is expressed through a very unfavourable sex ratio, high maternal mortality ratio and also high fertility rate. In most of these countries the sex ratio, that is, the number of females per one thousand males is not congenial to the women. A study of Table 13.4 reveals that this ratio is very ideal in case of Sri Lanka where in the total population the number of women and that of men are almost equal. However, this ratio is very poor in case of Pakistan, India and even Bangladesh while this ratio is a bit higher in Nepal. It is surprising to note that in India this ratio fell to 927 in 1991 from 934 attained a decade earlier. However, in 2001, this ratio moved again to 933. An unfavourable sex ratio is one of the basic symptoms of poverty of a country concerned. It is this acute poverty that makes a certain amount of bias in the allocation of many essential items, viz, food, health services, medicines etc. Again, it is because of this poverty that girls in many of these countries are tolerated rather than welcome– they are looked upon as unwanted children.

Table 13.4: Sex Ratio, Maternal Mortality Ratio and Total Fertility Rate in the SAARC Countries

Countries	Sex Ratio 1991	Maternal mortality ratio reported per 100000 live births, 1990-98	Total fertility rate 1991
India	927	410	3.9
Bangladesh	940	440	4.9
Pakistan	920	-	5.7
Nepal	950	540	5.5
Sri Lanka	990	60	2.5

Source: World Development Report, Various Issues.

One likely consequence of extreme poverty is the early marriage of the girls, the possible effects of which manifest themselves in the relatively large number of deaths among women in the reproductive age of 15-49, due partly to physical incapability to bear the burden of motherhood. This, along with extremely poor or non-existent medical facilities in remote areas and absence of pre-natal and post-natal care, accounts for high maternal mortality ratio as revealed in Table 13.4. This ratio is extremely high in Nepal, Bangladesh and India. The position of Sri Lanka is far better in this respect.

Acute poverty, high infant mortality rate and strong desire for having male children have led to an enhancement of total fertility rate (the average number of children that would be born alive to a woman during her lifetime if she were to bear children at each stage in accordance with the prevailing age-specific fertility rates) in most of the SAARC countries. Table 13.4 itself corroborates this phenomenon, from which it is seen that this rate is more or less reasonable in case of Sri Lanka. However, this rate is very high in Pakistan, Nepal, Bangladesh and India.

D. High adult illiteracy rate

Because of poverty most of the people do not have any urge for making themselves literate. Each and every person of a poor family is busy with earning something in either cash or kind to have a square meal a day. Parents get inclined to make their sons and daughters engaged in various jobs (sometimes illegal also) and they are least interested in giving them even the primary education. Themselves being illiterate and themselves being brought up in a poor atmosphere with poverty as the constant companion, the poor parents have no other alternative but to make their children illiterate. Table 13.5 will highlight the level of adult illiteracy in the SAARC countries.

It is clear from Table 13.5 that in the dissemination of education Sri Lanka has attained a marvellous success–only 7 percent of adult males and 16 percent of adult females are illiterate. In this respect Nepal is the worst sufferer, the position of Pakistan and Bangladesh is almost identical while India is slightly better. This table also demonstrates that illiteracy is more acute for females. It is a fact that poverty is the root cause behind massive illiteracy in most of these countries. It is simultaneously true that because of illiteracy the poor people fail to get their lot improved to a good extent.

Table 13.5: Adult illiteracy rate in the SAARC Countries

Countries	Adult illiteracy rate in percentage in 1990	
	Male	Female
India	38	66
Bangladesh	53	78
Pakistan	53	79
Nepal	62	87
Sri Lanka	07	16

Source: World Development Report, Various Issues.

E. Drinking water, health services and sanitation inaccessible to large population

A clear symptom of poverty in the SAARC countries lies with the fact that safe drinking water, health services and sanitation are still inaccessible to a large number of people. Table 13.6 will illuminate this phenomenon.

Table 13.6: Supply of drinking water, health services and sanitation in the SAARC countries

Countries	Percentage of population without access to		
	(a) Safe Water (1990-98)	(b) Health Services (1981-93)	(c) Sanitation (1990-98)
India	19	25	71
Bangladesh	5	26	57
Pakistan	21	15	44
Nepal	29	90	84
Sri Lanka	43	10	37

Source: Human Development Report, Various Issues.

Table 13.6 prominently demonstrates that in Sri Lanka and Nepal the respective percentage of total population still unable to get safe drinking water, is 43 and 29. In this respect Bangladesh has attained an enviable success while the position of India and Pakistan is almost identical. It is surprising to note that in Nepal as high as 90 percent of total population has no access to health services. In this respect the position of Sri Lanka is the best. Pakistan has also attained a good success while the position of India and Bangladesh is almost same. Coming to the access to sanitation we see that Nepal is the worst sufferer while the position of Sri Lanka is far better. In this respect India has fared very badly.

CONCLUSION

Judged by the above-mentioned criteria of the prevalence of poverty in the SAARC countries we see that the economic scenario prevailing in this region differs from country to country. It is because of this reason that the countries comprising the SAARC have attained different Human Development Indices that have been presented in Table 13.7.

Table 13.7: Human Development Indices in the SAARC countries

Countries	Human Development Index, 2000
India	0.577
Bangladesh	0.478
Pakistan	0.499
Nepal	0.490
Sri Lanka	0.741

Source: Human Development Report, 2002, Oxford.

From Table 13.7 we notice that the highest Human Development index has been attained by Sri Lanka. Next to Sri Lanka this achievement has been shared by India. In this race the position of the remaining three countries is almost identical, though Pakistan has an edge on both Nepal and Bangladesh and Nepal has that on Bangladesh.

The basic factor leading to massive poverty in the SAARC countries is the existing population together with the growth rate of population. It is really a matter of grave concern that the total fertility rate of each of these countries is very high. It is because of this high fertility rate that population in these countries has been increasing. Being unable to effectively utilizing this human resource these countries have been facing the problem of abject poverty.

Policies prescribed for eradicating poverty

To ameliorate poverty from most of the countries referred to above we may prescribe the following policies:

1. Rapid growth of population in the rural areas is the key causal factor leading to massive poverty. The population growth is larger among the poor. Prof. G. Ackley in his Macro Economic Theory has rightly pointed, "Low income families include many victims of temporary bad luck." The faster growth of population among the poor caused partly by poverty aggravates it instead of eliminating the same. This calls for a reduction in the steep rise in the growth rate of rural population. For this the modern devices of birth control should reach at the doorstep of the rural poor and they should be convinced of unhesitatingly utilizing the same.
2. Education should be disseminated in the far and near of the countries concerned. Every child should take at least primary education and in this respect there must not be any discrimination between caste, creed and sex. The girl children should no longer be neglected and they should be brought up in the same manner as the boys are done. It is a fact that there is an inverse relationship between the attainment of education and the total fertility rate. Besides, an educated mother is more productive in the maintenance of the family than an educated father. Thus with the spread of education, specially among the women, the fertility rate will come down and thereby the growth rate of population will be checked and this will lead to a reduction of poverty.

3. A very peculiar character of some class of rural poor is that they do not generally care for what is going to happen in near future. It is because of this nature that they fail to take the opportunity of earning more or even maintaining the spirit of earning through different jobs. Such nature keeps them in the poverty trap from which they cannot, easily, come back. They should be given proper training and direction in this respect.
4. There should be some reservation of development in favour of the weaker sections of the society. They should have a share in the additional assets and incomes generated through public expenditure and development programmes.

Together with the above-mentioned policies we should also recall the dictum that only through toil and trouble, sweat and blood, hard work, honesty and efficiency and proper distribution of fruits of development should we be able to attain higher standard of living and greater welfare.

REFERENCES

1. Ahluwalia, Montek (1978): Poverty in India and in the States, *Journal of Development Studies.*
2. Bhattacharyya, N., Coondoo, D, Maiti, P., and Mukherjee, R. (1991): *Poverty, Inequality and Prices in Rural India*, Sage Publications.
3. Census of India, Various Issues.
4. *Human Development Report*, Various Issues, Oxford.
5. Konar, D. N. (1996): "Problem of Food and Malnutrition in the SAARC Countries", *Artha Beekshan*, Vol. 5 No.2 Dec.
6. Konar, D. N. (1998): "Unfavourable Sex Ratio and the Deprivation of Women in India", *Artha Beekshan*, Vol. 7 No. 2 Dec.
7. Planning Commission (1993): *The Report of the Expert Group on Estimation of Proportion and Number of Poor*, Perspective Planning Division, Government of India, New Delhi.
8. *World Development Report*, Various Issues, World Bank.

Conclusion

In Economic literature a relatively new term called "Human Development" launched in 1990 by the noted economist late Professor Mahbub-ul Haq has been widely accepted. The Human Development Report has defined this term as the process of (a) enhancing the basic capabilities and (b) enlarging the range of choices by the people such that they will be able to improve the means of their well-being. Besides a high level of income, a long and healthy life, upgradation of knowledge and skills, gender equality, social awareness and initiative of people, people's participation in administration etc. are critical factors for development of a nation.

The information supplied and the analysis made in most of the chapters of the present work make it clear that in the field of human development our progress has not been satisfactory and a lot remains to be accomplished. Two very vital forces of development, namely women and children, have not been given proper attention. The literacy rate among women in our States and even in SAARC countries, is not yet satisfactory. Women are still deprived of many things—we still hanker after having at least a male child (not a girl one) and naturally sex ratio is going downward and downward. This ratio in the age group 0 to 6 is really a matter of concern. In the society the abortion of female foetuses is going on unabated; women are still not paid due status, honour and respect. The declining sex ratio among children will have a serious impact upon the overall sex ratio of the country and also of the states when the result of the coming census of 2011 will find the light of the day. Similar is the case with numberless children who are forced to wither before being adults. The natural consequences of all such maladies have become severe in our family life, in our social life and our country as whole. As a result of nutritional deprivation, especially among girls, more than 30% of the children are born underweight in our country. The picture is more or less identical in most countries belonging to SAARC. Many children are born blind and handicapped. Malnourished and underweight children are evidently a burden to the society.

It is true that the present children are the future of the nation. But what can be expected of the malnourished and underweight children? It is a fact that poverty is the root cause behind all such undesirable outcomes. But besides poverty, a high rate of growth of population is also held responsible for such a sorry state of affair. We have not yet been able to make primary education compulsory and to reduce the growth rate of population to the desirable extent through family planning programme for the introduction of which India is a pioneer. Indonesia known as a 'hub' of Asia's economic development, has been very successful in implementing the family planning programme and it has been able to lower the growth rate of its population. Naturally, Indonesia has attained a higher growth rate though both India and Indonesia had attained independence at about the same time. Needless to say, as an honest and bona-fide citizen of this great country we have a lot to do in this direction.

The philanthropist and the noted Professor of both Economics and Philosophy, the Nobel laureate Dr. Amartya Sen has been, for a number of years, teaching us to give more emphasis on human development. However, the root of this approach can be traced back about two hundred years earlier to a song of a Bengalee Saint Ramprasad who used to live at Halisahar (about 40 km. away from Kolkata) near Kalyani town in West Bengali: "A lot of human land remains fallow, which could have produced gold had that been cultivated." The terrible and man-created Bengal famine of the forties of the last century had made life of most people of the then Bengal extremely miserable and people's sorrow, grief and agony knew no bounds. Such pain, anxiety and intolerance had given a rude shock to the "child Amartya" who had cherished the painful and hungry scene of the human being in his mind and subsequently he taught the whole world to think over the welfare of the people more realistically and more practically. He has taught us to give more importance to primary education, health care and to enhance the dignity and status of women who comprise about half of total population of the world.

Proper attention should also be paid to the senior citizens of our country. They are not burden to the society. In fact, many of them can be made assets of the society. With due regard, respect and honour we can get a lot from most of them.

The very basis of the Indian economy is, still today, villages that have not been adequately taken care of. As a result, people have been leaving villages and have been searching for residence in the urban areas. Naturally, these areas are becoming overcrowded. An obvious impact of this is the generation of various types of pollution and different types of problems, including water problem, space problem, problem arising due to non-co-operation among members of the same co-operative societies. Needles to say, life in the urban areas is also becoming miserable. To get rid of such outcome we suggest that the rural areas should be provided with some minimum amenities, otherwise it will be quite difficult to protect the urban areas from the encroachment of the migratory people.

In fine, we would also say that though the trend growth rate of population in India and many States has declined, it has not yet come to the desirable figure. For reaching this figure, serious attempts are to be made by people of all walks of life. At the same time, the declining trend of birth of girl children should, immediately, be arrested and all persons found guilty in this regard should be severely punished.

Appendix

A. Fitting of the Exponential Curve and the Determination of the Annual Compound Rate of Growth.

Throughout our analysis by growth rate we have meant the annual compound or exponential or trend rate of growth. This rate of growth can be found by fitting the exponential curve having the form

$$Y = ab^t \qquad (1)$$

where t is time, y is the relevant variable (say, population over different years), a and b are two parameters whose values are to be determined from the given time series.

Before dwelling on the process of fitting the curve represented by equation (1) let us narrate, in brief, the fitting of the straight line or linear trend having the form

$$Y = a+bt \qquad (2)$$

The parameters involved in equation (2), that is, a and b can be found by solving the following two normal equations:

$$\Sigma y = na+b\Sigma t \qquad (i)$$

and

$$\Sigma y \, . \, t = a\Sigma t + b\Sigma t^2 \qquad (ii)$$

Here n is the number of years involved in the given time series. From the given time series through the appropriate choice of origin and scale t may be selected in such a manner that the sum of odd powers of t (that is, Σt, Σt^3, Σt^5, . . .) may be made equal to zero. Hence the above two normal equations may be rewritten as

$$\Sigma y = na \qquad (i^1)$$

and

$$\Sigma y \, . \, t = b\Sigma t^2 \qquad (ii^1)$$

$$\text{Thus } a = \frac{\Sigma y}{n} \text{ and } b = \frac{\Sigma yt}{\Sigma t^2}$$

These calculated values of a and b put in equation (2) will give the necessary straight line.

Let us now turn to the fitting of the curve represented by expression (1). This curve is, obviously, non-linear. However, it is log linear as

$$\log y = a + t \log b \qquad (i^1)$$

$$\text{or, } y = A + Bt \qquad (ii^1)$$

Since the expression (i^1) is linear in form, its fitting will be similar to the fitting of (2). Hence the values of log a and log b can be found by solving the normal equation:

$$\Sigma \log y = n \log a + \log b \ \Sigma t \qquad (iii)$$

and

$$\Sigma \log y.t = \log a \ \Sigma t + \log b \ \Sigma t^2 \qquad (iv)$$

Also by previous technique $\Sigma t = 0$. Hence we have

$$\Sigma \log y = n \log a \qquad (iii^1)$$

and

$$\Sigma \log y.t = \log b \ \Sigma t^2 \qquad (iv^1)$$

$$\text{That is, } \log a = \frac{\Sigma \log y}{n} \text{ and } \log b = \frac{\Sigma \log yt}{\Sigma t^2}.$$

By taking anti-log we can, at ease, find out the values of a and b.

A scientific calculator may be of much help for finding out the values of a and b.

Let us now see how the annual compound (or trend) rate of growth can be found.

We know that a sum of money M_0 invested for t years at r% compound rate of interest will accumulate to S_t where

$$S_t = So \left[1 + \frac{r}{100}\right]^t \qquad (3)$$

Comparing equation (1) and (3) we see that

$$b = 1 + \frac{r}{100}$$

where

$$\frac{r}{100} = \text{b-1}.$$

Thus r, the annual compound rate of growth, will be (b-1) × 100%.

The readers keenly interested in fitting other mathematical curves, namely modified exponential curve, gompertz curve and the logistic curve, are suggested to kindly go through the references.

B. Rank Correlation Coefficient and its Use

In Social-science research the rank correlation coefficient is very often used.

When the two variables are quantitatively measurable, the degree of linear interdependence between them can be measured by calculating Pearson's product moment formula. However, cases many arise in practice when we are to deal with attributes which are not expressible as definite numbers directly. It is possible, in such cases, to rank the objects with regard to the quality. For instance, we can rank a group of persons with regard to their intelligence and efficiency in salesmanship, using numbers 1, 2, 3 . . . The individuals are then said to be ranked and the number allotted to a particular person is called its rank. For each individual, we then have a pair of ranks, one for each character. If x and y represent ranks of the same individual or object with regard to two different qualities, the product-moment coefficient of correlation between x and y will be called the rank correlation coefficient which is measured by

$$r_R = 1 - \frac{6\Sigma di^2}{n(n^2 - 1)},$$

where di = (xi - yi), n being the number of individuals to whom ranks are allotted.

It can be shown that

$$r_R = \begin{cases} +1 \text{ in case there is perfect agreement between the two series of ranks} \\ -1 \text{ in case of perfect disagreement between the two series of ranks.} \end{cases}$$

Let us now consider the case of tied ranks. If the same rank is allotted to 2 (or 3 or more) individuals, then we have a tie of length 2 (or 3 or more). In dealing with such cases, the usual way is to allot the average rank (that is, the arithmetic mean of the ranks that these individuals would have received had there been no ties) and then calculate the product-moment coefficient of correlation from these ranks.

The serious readers many go through any good book on Descriptive Statistics dealing with Correlation Analysis.

REFERENCES

1. Croxton, F.E. and Cowden, D.J., (1967): *Applied General Statistics*, Prentice-Hall, (chs. 11-14, 16).
2. Goon, Gupta and Dasgupta (2000): *Fundamentals of Statistics*, Vol. I and Vol. II, The World Press Pvt. Ltd., Kolkata.
3. Kendall, M.G. (1948): *Rank Correlation Methods*, Charles Griffin, (chs. 1-3).
4. Kendall, M.G. and Stuart, A. (1958): Advanced Theory of Statistics, Vol. I, Charles Griffin.
5. Konar, D.N., (1997): "Trends of Indian Agriculture in the Post-Independence Period", *Artha Beekhsan*, Vol. 6, No. 2.
6. Mills, F.C. (1995): *Statistical Methods*, H. Holt.
7. Yule, G.U. and Kendall, M.G. (1950): *An Introduction to the Theory of Statistics*, Charles Griffin.

Select Bibliography

Ackley, G: Macro-economic Theory (old edition).

Banerji, Anita and Sen, R.K. (2000): Women and Economic Development, Deep & Deep Publications Pvt. Ltd., New Delhi.

Bose, Ashish (1991): Population of India - 1991 Census, Results and Methodology, April.

Census of India, Various Issues.

Coale, A.J. and Hoover, F.M.: Population Growth and Economic Development in Low-Income Countries.

Colin Clark: Conditions of Economic Progress.

Datt, R and Sundharam, K. P.M. (2002): Indian Economy, S. Chand & Company Limited, New Delhi.

Datt, R. and Konar, D.N. (2000): Social Choice and Development, Deep & Deep Publications Pvt. Ltd., New Delhi.

Dutta, Bhabatosh: Economics of Industrialization.

Economic Review, Govt.of West Bengal, Various Issues.

Economic Survey, Govt. of India, Various Issues.

Giri, Pabitra (1998): "Urbanization in West Bengal, 1951-1991", Economic and Political Weekly, Vol. XXXIII No.s 47 and 48.

Goon, A.M. and Halder, D. (1990): "Calcutta's Population: Changing Size and Age Sex Composition", Arthashastra, Vol. 1, No. 2.

Goon, Gupta and Dasgupta (2000): Fundamentals of Statisties, Vol. I and Vol. II, The World Press Pvt. Ltd., Kolkata.

Hajela, P.D. (1996): "Education as a Tool for Social Uplift", Yojana, Vol. 40, No. 1, January.

Human Development in South Asia, 1997.

Human Development Report, Various Issues.

Indian Development Report.

Kannan, K.P. (ed) (2001): Economics of Child Labour, Deep & Deep Publications Pvt. Ltd., New Delhi.

Konar, D.N. (1988): "Changing Senario of Indian Population: A Projection to 2000 A.D.", Artha Shastra, Vol. 7 No. 1.

Konar, D.N. (1999): "Trend of Urbanization in West Bengal Since Independence", Artha Beekshan, Vol. 8 No. 1.

Konar, D.N. (1998): "Population Pressure and the Problem of Environment in Calcutta", Journal of Commerce, Vidyasagar University, Vol. 3, March.

Konar, D.N. (1996): "Urbanization in West Bengal and the Problem of Environment", Economic Affairs, Vol. 41 No. 1, January-March.

Misra, L (2000): Child Labour in India, Oxford University Press, New Delhi.

Monga, G.S. (2001): Environment and Development, Deep & Deep Publications Pvt. Ltd., New Delhi.

Munshi, S.K. (1975): Calcutta Metropolitan Explosion, People's Publishing House.

Sen, Amartya (1982): "Poverty and Famine: An Essay on Entitlements and Deprivation", Oxford Unviersity Press, New Delhi.

Sen, Amartya (2000): Development as Freedom, Oxford University Press.

Sen, R.K. (2002): Indian Labour in the Post-Liberalisation Period, K.P. Bagchi & Company, Kolkata.

Statistical Outline of India (2001-02, 2002-03): Tata Services Limited.

Visaria, P and Visaria, L: "Indias Population after 1981 Census", Economic and Political Weakly, Vol. XVI, No.s 44, 45 and 46, Special Number.

Todero, M. (1987): Economic Development in the Third World, Orient Longman.

The World Development Report, Various Issues.

UNICEF (1997): The State of World's Children, Oxford University Press.

Index